Plan Bee

Everything You Ever Wanted to Know About the Hardest-Working Creatures on the Planet

SUSAN BRACKNEY

HAY HOUSE

Australia • Canada • Hong Kong • India
South Africa • United Kingdom • United States

First published by Perigree, an imprint of the Penguin Group (USA) Inc.
375 Hudson Street, New York, New York 10014, USA
Penguin Books Ltd. Registered Offices: 80 Strand, London WC2R 0RL, England

First published and distributed in the United Kingdom by:
Hay House UK Ltd, 292B Kensal Rd, London W10 5BE. Tel.: (44) 20 8962 1230;
Fax: (44) 20 8962 1239. www.hayhouse.co.uk

Published and distributed in Australia by:
Hay House Australia Ltd, 18/36 Ralph St, Alexandria NSW 2015. Tel.: (61) 2 9669 4299; Fax: (61) 2 9669 4144. www.hayhouse.com.au

Published and distributed in the Republic of South Africa by:
Hay House SA (Pty), Ltd, PO Box 990, Witkoppen 2068. Tel./Fax: (27) 11 467 8904.
www.hayhouse.co.za

Published and distributed in India by:
Hay House Publishers India, Muskaan Complex, Plot No.3, B-2, Vasant Kunj,
New Delhi – 110 070. Tel.: (91) 11 4176 1620; Fax: (91) 11 4176 1630.
www.hayhouse.co.in

Text design by Tiffany Estreicher

A catalogue record for this book is available from the British Library.

ISBN 978-1-8485-0193-5

Printed and bound in Great Britain by CPI Bookmarque, Croydon, CR0 4TD.

CONTENTS

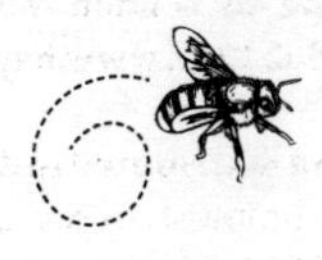

The happiness of the bee and the dolphin is to exist.
For man it is to know that and to wonder at it.
—Jacques Cousteau

INTRODUCTION

The Birds and the Bees

This very nearly could've been a book about chickens. See, I'm big on self-reliance, and I also happen to have a soft spot for soft creatures. So, after countless trips to the "Rabbit and Poultry" building at the county fair, I was set on having my own flock of hens. And not just any hens. I would have absurdly fancy ones like the Polish silver lace. Its bouffant of black-and-silver feathers looks like a lady's expensive, over-wrought hat. Or I might choose those rather overdressed, buff-colored Cochins wearing their unruly, feathered pants and matching, feathery shoes. Not only would I collect their eggs, but I planned to sneak out to pet them from time to time as well.

To realize my dream, I bought a tiny house on a half-acre lot just outside of town, but the ink was hardly dry on the paperwork when the city annexed my little patch of para-dise. I'd never paid much attention to the goings-on of local

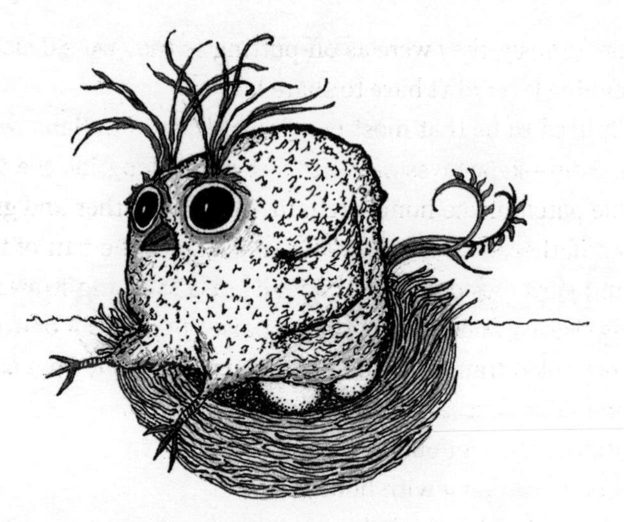

government back then, and as such, I paid the price—citified chickens were strictly prohibited.[1] That was the start of a series of events that, once set into motion, would turn my chickens into honeybees.

* * *

Eventually, I adjusted to life without chickens. Having an enormous garden helped. I sought out the old-timey vegetable varieties that my great-grandmother and her mother and even her mother's mother might've grown. I grew scads of snap beans, black raspberries, and greens. There were zinnias as tall as Danny DeVito and, enormous and purple, the heirloom tomatoes, which, when sliced, looked a lot like raw beef.

1 Ah, but no longer. I am proud to say that our city fathers eventually agreed to allow residents to have up to five hens, as long as one's surrounding property owners approve.

(Fortuitously, they were as off-putting as they were delicious, meaning I wouldn't have to share.)

It used to be that most people—at least in Indiana, where I'm from—kept hives of honeybees right alongside the vegetable patch or the home orchard. My grandfather and great-grandfather did. I have a 1930s snapshot of the pair of them leaning over a small hive amid tall grass. Wearing straw hats, long sleeves, and overalls, they're poised to grab out one of the honey-filled frames. When he was just a boy, my own father, some fifty feet away, would watch as they cut out sections of comb, dripping with honey, to bring inside. Fresh honey was a given then, and beekeeping wasn't yet a dying art.

Howard Goodin Sr. and son check their bees. *Author's family photo*

Partly to pay homage to that more self-sufficient time, I thought I might try beekeeping myself someday. A hive of bees would likely pollinate the heck out of my fruits and vegetables, I reasoned. Besides that, I'd heard that honeybees were running out of natural habitat and that managed honeybees were losing their graying keepers, who were passing away before passing on their specialized knowledge. I took a short night class on beekeeping to see what it was all about. That's when I realized I'd need quite a lot of expensive equipment to get started. Between the city's red tape and my modest means, it looked like I'd have neither chickens nor bees.

* * *

We never really know what impression our words have on others, but apparently, something I said about wanting to keep bees really stuck with my friend Michael. Lucky for me, he is both very generous and compulsively drawn to yard sales. Luckier still, a beekeeper I'd never met fell deeply in love with a lady who was allergic to bees. The pair got married, and not long after, he held a yard sale that included his hives, honey extractor, smoker, veils, gloves, and back issues of *Bee Culture* magazine.[2] It was all easily worth thousands of dollars, but he sold the whole lot to Michael—who happened by at just the right time—for a song. I was puttering out in my garden when Michael called me with the exciting news:

MICHAEL: *Hey, do you still want to have honeybees?*
ME: *Yeah . . . Why?*
MICHAEL: *Because you're a beekeeper now!*

Set aside the infinite spoonfuls of honey, the batches of homemade mead, and so many sweet-smelling beeswax candles and soaps, and I can still say the honeybees have brought

2 I am still pleasantly surprised that, in this age of divorce, a newly married beekeeper would sell all his gear rather than put it in storage just in case the marriage should fail. Though they don't really know me, I often think about the happy couple and the faith embodied in the man's gesture.

an extraordinary sweetness to my life. As I idly watch them fly from flower to flower and from flower to hive, I realize they've managed to slow my previously frenetic pace, to make me more appreciative of the workings of the universe, and to return me, at least a little bit, to simpler times. As you get to know the fascinating, persevering, life-affirming honeybee, I hope all these things for you, too.

Part One

The Buzz About Bees

CHAPTER ONE

On Being a Bee

Honeybee Impostors

I am, at a moment's notice, ready to remove the odd swarm of wild honeybees dripping from some old lady's porch light. As such, the local animal control department has me on speed dial, and each time I'm called, I happily pile my bee veil, spare hive boxes, and bee smoker in the trunk of my car and get going. But I can't tell you how many times I've arrived at someone's house expecting a bounty of bees clinging to that porch light or some high tree branch, only to find a stream of yellow jackets pouring out of a shiny hole in the ground. Each occurrence is bitterly disappointing. Fortunately, I've wised up. Now I ask a few key questions before I bother to come out. Among them: Do the insects you have look really shiny with bright yellow and black? Are they nesting in the ground? Do they seem rather aggressive? And when people answer, "Yes,

704

yes, and yes!" I give them the bad news. Probably just yellow jackets. "But they're stripey!" they often protest. Um. It takes more than stripes to make a honeybee, honey.

To set the record straight, yes, both honeybees and yellow jackets have stripes, but honeybees have a much greater color range—some of my bees are nearly all black, others are caramel-colored, and still others are a brilliant yellow-gold. A honeybee's abdomen does look shiny, but its thorax is somewhat fuzzy. Finally, while yellow jackets scavenge around garbage cans for meals, the gentle honeybee painstakingly collects pollen and nectar for herself one magnificent flower at a time. Still, we haven't had quite as many honeybees flying around as we used to, so I can see how most people can't exactly pick one out of a lineup of, say, a yellow jacket, a bald-faced hornet, and a bumblebee.[1] I can see it now: Looming large, the praying mantis guard unfolds her hands and motions the lineup's subjects to turn to the side. The yellow jacket sneers a little and then complies. The bald-faced hornet methodically chews up her paper number instead. The bumblebee, likewise, remains in place, entirely distracted by the sound of her own buzzing. As for our poor honeybee? She's overwhelmed by the feeling that she has better things to do. Oh, but she is used to these cases of mistaken identity. Her irascible cousin, the yellow jacket—along with the hornet, the

1 In the last couple of years, some commercial beekeepers have lost more than half of their hives to Colony Collapse Disorder, a mysterious malady researchers are still trying to understand. Shrinking natural habitat has also contributed to a dearth of wild honeybees.

bumblebee, and a host of ants and sawflies—belongs to the same scientific order, the Hymenoptera.[2]

The Bee Family Tree

My family tree is small; some branches have cracked, and others are missing altogether. Needless to say, I was never too fascinated by genealogy, but I think I could've been—had I been born a honeybee, that is. Generally thought to have descended from wasps, there are all sorts of bees, including one so tiny that it barely fits on the head of a pin and still another that's golf ball–sized with a 2.5-inch wingspan. There are the "killer" bees willing to chase nest invaders for as far as a quarter mile, and there are bees that cannot sting but, instead, will *bite*. Despite their differences, all these bees are lumped together as members of the Apoidea superfamily, which is then subdivided to include bumblebees, stingless bees, and honeybees, among others.[3] Now, the people who've made it their business to classify the world's flora and fauna include honeybees in the genus *Apis*—that's Latin for "bee." The common honeybee is known as *Apis mellifera*, which translates to "honey-bearing bee," and the *Apis mellifera* portion of the bee family tree splits off in some pretty interesting directions.

Like each of us, *Apis mellifera* has been shaped by the "neigh-

2 Derived from the Greek roots, *hymen*, or "membrane," and *ptera*, which means "wings," Hymenoptera's members have clear-membraned wings.

3 Stingless bees live in tropical and subtropical areas including Africa, Australia, and South America, and although they may sound harmless, some types of stingless bees can be surprisingly aggressive.

borhoods" in which she grew up. The honeybees that originated in Europe, Africa, and the Near East can exhibit vastly different habits and temperaments, characterizing the many subspecies or "races" of *Apis mellifera*. For one, hailing from northern Europe and parts of Russia, *Apis mellifera mellifera*, or the German dark bee, was said to tolerate harsh winters well. As her name suggests, the German dark bee had near-black markings. Despite her rather aggressive disposition, she was likely the first honeybee introduced to North America.[4] It wouldn't be long, though, before *Apis mellifera ligustica*—a much gentler, golden bee from Italy—crowded her out. One of my favorite beekeeping collectibles is my 1917 edition of A. I. and E. R. Root's *The ABC and XYZ of Bee Culture* which didn't mince words when it came to the value of German versus Italian honeybee stock:

Trapped in amber, the oldest fossilized bee ever discovered dates back 100 million years.

> *[German dark bees] are much more nervous; and when a hive of them is opened they run like a flock of sheep from one corner of the hive to another, boiling over in confusion, hanging in clusters from one corner of the frame as it is held up, and finally falling off in bunches to the ground, where they continue a wild scramble in every direction, probably crawling up one's trouser leg, if the opportunity offers.*

I have worked with both the "dark" European bees as well as the "light" Italian kind, and I admit, the Italians did seem less eager to sting, But to the Roots writing back in 1917, Ital-

4 Honeybees were not native to North America, and the Native Americans took to calling honeybees the "White man's fly."

ian bees could simply do no wrong: "At present the Italians, and even hybrids, have shown themselves so far ahead of the common bee that we may safely consider all discussion of the matter at an end by the majority of beekeepers."

German bees, in contrast, could do no right. World War I still raged at the time, and I think that fact surely helped to color the authors' somewhat bizarre perceptions. Of the German dark bees they write:

> *Their queens are much harder to find, the bees are not so gentle; and, worse than all, they have a disagreeable fashion during robbing time of following the apiarist about from hive to hive in a most tantalizing manner.*[5] *This habit of poising on the wing in a threatening manner before one's eyes is extremely annoying, and some bees will keep at it for a day at a time unless killed. We generally make very short work by smashing them between the palms of our hands, or batting them to death with little paddles that we keep near.*

Besides those beloved Italians, U.S. beekeepers also gravitated to *Apis mellifera carnica*, or the Carniolan honeybee, which originated in Slovenia. The Carniolan is a dark-colored bee, like *Apis mellifera mellifera*, but she is said to be one of the gentlest of all the honeybee subspecies and has long been recommended for beginning beekeepers. Another kindly, dark gray bee that some U.S. apiarists like is *Apis mellifera*

5 I've never really cared for the old-timey expression "robbing the bees," since I don't like to think of myself as a honey thief. But that's how the old timers often looked at beekeeping and their role in it.

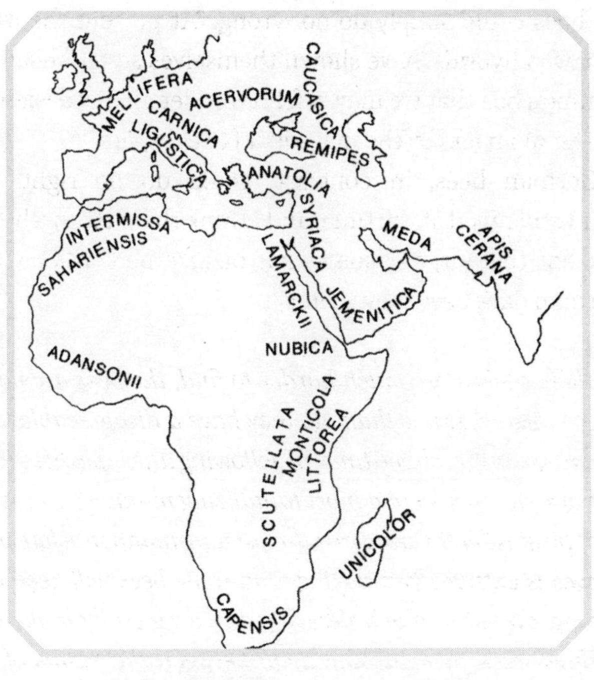

Where it all began for the many honeybee subspecies. ***Harvard University Press***

caucasica, the Caucasian honeybee. *Apis mellifera caucasica* originally came from Eastern Europe's Caucasus Mountains region and can be a little harder to find.

Oops . . .

Not every honeybee is nearly so mild-mannered. Aggressive enough to kill livestock, people, and pets, one African subspecies, *Apis mellifera scutellata*, has been the gentler bees' public relations nightmare since the late 1950s. It started innocently

Beekeepers in rural Venezuela wear heavy-duty veils to protect their faces while inspecting the hives of some excitable Africanized bees. *James E. Tew, The Ohio State University*

enough when, in 1956, Brazilian entomologist Warwick Kerr won a government grant to build a hardier honeybee. Kerr thought shoring up the beekeeping industry could help his country's poor, so he tried to create a Brazilian-African hybrid that could better withstand Brazil's tropical temperatures. To that end, scores of *Apis mellifera scutellata* queens were shipped to Brazil, but in 1957, twenty-six of them were accidentally released into the wild. That might not have been such a big deal—if *Apis mellifera scutellata* weren't so touchy.[6] While Africanized honeybees look like their European counterparts,

6 An easy way to remember the "scutellata" part of *Apis mellifera scutellata*: If you see Africanized bees coming to attack, you'd better *scoot a lotta* tail out of there! Very bad pun, I know, but that is how I remember *Apis mellifera scutellata* to this day.

they certainly don't act like them. If they perceive that their nest is being threatened, Africanized bees will attack by the hundreds rather than the dozens, and they'll fly much farther to mete out their punishment.[7]

Not that I know that from personal experience. Truly, hearing an Africanized honeybee attack on the radio is the closest I've come to understanding just how thin-skinned *Apis mellifera scutellata* can be.

As I recall, a honeybee expert agreed to have a radio soundman accompany him into an apiary populated with Africanized honeybees. Wearing protective suits and veils, the men stepped in range of the hives, and in just a few seconds, they were engulfed by angry bees. There had been so many bees clinging to the men's veils that it was difficult for them to see, and I'd marveled at the "How dare you?!" tone I detected in the bees' collective buzz. In short order the bees managed to sting through all that protective gear, and they were in such a frenzy that the air had filled with the tell-tale scent of banana.[8] I confess I was relieved to be safely on the other side of the radio.

Over the years Africanized bees have thrived, completely displacing European

7 Africanized honeybee venom is no more dangerous than that of European bees, but anyone attacked by Africanized bees is likely to sustain many more stings—and more stings means more bee venom.

8 Some people swear they can smell when the bees are upset. Alarm pheromone reportedly gives off a fruity odor, but I've never smelled it myself.

bees in some regions of the Southwestern United States. Now you can find them in the six Southwestern states from Texas to California and in the Southeast from Arkansas and Louisiana to Florida. Dewey Caron, a professor of entomology and wildlife ecology at the University of Delaware, says someday they may also live in Northern states like mine: "They are better suited to subtropical conditions, where they thrive, but as they move into areas that have more temperate climates and more severe winters, they adapt." In subtropical areas, Africanized bees displace the native bee populations in just a few years. In more temperate areas, though, the process might take a decade or two. That's not good news, but Africanized bees aren't *impossible* to keep.[9] Just a bit different, as Caron explains, "You always have to have the very hot, heavy protective equipment even on 100-plus-degree days. . . . You have to plan for their being more aggressive, and you have to plan for the possibility that an accident may occur." By now, South American beekeepers have adjusted, and one day, if I must, I will, too.

9 I met a beekeeper in Arizona who tends hives of Africanized bees, and he said it's not so bad. He's able to charge a premium for his "killer" bee honey—even though it's no different than the sweet stuff a European bee would make. And I'm told the local ladies think he's something of a badass.

CHAPTER TWO

Who's Who in the Hive

Girl Power

Whether they're Africanized bees with a short fuse or the mellowest of Italians, one thing remains the same in every beehive: Female bees run the show. For starters, almost as soon as the all-female worker bees emerge from their brood cells, they set about to cleaning and polishing other honeycomb cells, so that they can be used again. And they busy themselves feeding the surrounding baby bees, which seems noble enough, but the really good jobs go to the older workers. They shape new beeswax cells, help to store any nectar and pollen brought in by other, foraging worker bees, and in due course, they get to take their own foraging excursions. Worker bees also spend some time ripening honey and cooling the hive on very hot days.[1]

1 In a sense you could say honeybees have perfected their own kind of air-conditioning. When hive temperatures reach about 95 degrees and up, some foraging

From left to right: the (female) worker bee, the queen bee, and the (male) drone.

Blocking ants, yellow jackets, and other invaders, worker bees eventually will guard the hive entrance, and some bees even serve as undertakers, keeping the hive tidy by unceremoniously hauling out the dead.[2] I imagine being a worker bee never gets dull, since each gets to rotate through nearly all the available hive duties at some point or other. (That way, if chewing beeswax isn't your thing, you'll be flying among so many foxglove and poppies soon enough.) There is at least one other job for a worker, which may well be her highest calling, and that's tending

---

workers will collect water to spread around inside the hive, and then several rows of bees—some inside the hive and others just outside its entrance—will fan their wings simultaneously to create a cooling draft. Bees also beat the heat by sticking themselves on the front of the hive in one large mat or "beard." On very hot days some of my hives look as if they've been turned inside out, with more bees clinging to the outsides than in.

2 Sometimes just one bee will stagger out of the hive entrance, lugging a corpse as best as she can, but more often I've seen bees working in pairs—one on either side of a body—to fly it as far from the hive as they're able.

to the queen. At any one time, about a dozen queen attendants groom, feed, and clean up after Her Highness, who is simply too busy to be bothered with the day-to-day. Quite literally the life of the hive, when it's time to build up the numbers of her subjects in the spring, the queen will lay between 1,500 and 2,000 eggs *each day*.

A queen bee surrounds herself with devoted servants. *Kim Flottum*

I often try to spot her on the job, but that isn't easy. When I examine the lower reaches of the hive to see just how many baby bees are being produced, the queen runs like a starlet dodging the paparazzi. It's not that she's being coy. Actually, once she has mated, she prefers total darkness. Expose her cozy brood chamber on a sunny day, and it's just too much for her to bear. Occasionally, I've glimpsed her running away, her attendants in tow, and each time is exhilarating. I feel as if I've seen someone famous. Or at least, someone really important. My sexy Italian queen is slender and much longer than the workers surrounding her. Her caramel-colored abdomen brims with all the eggs and sperm she'll ever need. So many people have asked me how I can be sure when I've located her, and I often say, "She's the one wearing the tiny, gleaming crown." You wouldn't believe how many souls, all caught up in her glamour and mystique, say, *"Really?!"* No, not really, but it sure would be neat—and certainly convenient. In my early beekeeping days, I attended a live queen-finding demonstration in the hopes that my own queen would no longer be so able to elude me. Had you been there, you would've overheard something like this . . .

THE INSTRUCTOR: *There she is! Do you see her?*

THE REST OF THE CLASS: *Oooooh!*

ME: *No.*

THE INSTRUCTOR: *There! See?*

THE REST OF THE CLASS: *Ahhh! Wow!!*

ME: *No.*

THE INSTRUCTOR, POINTING IN EXASPERATION: *Well, she's* right there.

ME, FEIGNING EXCITEMENT: *Oh, huh. I see her now . . . sort of . . .*

I still have difficulty locating the queen, and now I look for the next best thing—evidence that she is well and laying egg after egg. They aren't too tiny to be noticed, anchored on one end to the bottom of the honeycomb cells in which they are cradled. Because the queen is always laying, there should be baby bees in several stages of development. If I can find neither the queen herself nor just-laid eggs, I assume the worst. Yes, some queens have lived for seven or even eight years, but, lasting a year, maybe two, few fare that well.

There is something downright brutal about a beehive, after all. Things can be buzzing right along, and then—blammo!—your queen gets it. Sometimes it's natural causes, and accidents do happen, but more often it's murder. Just as aging women in human populations seem to lose their worth—sometimes even their very *visibility*—so, too, does the elderly queen. Honeybees will turn on her if she's hurt or just not laying as well as she used to. In a process beekeepers call "supercedure," a cabal of worker bees chooses spots already containing eggs and turns the cells containing them into special cells fit for their new-and-improved queen.[3] They feed these queens-to-be nothing but royal jelly, seal the cells with beeswax, and wait.[4] A week or so later, several of the queens will emerge and fight one another to the death. The last queen standing wins, and if there are any queens still sealed up in their respective queen cells, she'll sting them to death through the cell cappings, making her ascension to the throne *almost* official. Before a new queen can get busy in the hive, she must get busy with some drones. The queen mates with several of them high in the air, during what are known as her

Because they're differently positioned and larger than regular honeybee cells, queen cells like this one really stand out.

3 "Queen" cells are much larger than regular brood cells, and rather than being placed in line with all the other cells on a frame, queen cells are placed vertically. It almost looks as if someone has affixed beeswaxy peanuts to the sides and along the bottom edges of the honeycomb.

4 Royal jelly is strange and powerful stuff, and you'll soon see what I mean.

"maiden" flights. Successfully mated, the queen returns to the hive with over five million sperm stored up inside her. She'll begin laying her fertilized eggs within a few days, at which point, if the hive's *old* queen has been allowed to live this long, either the new queen or the old queen's once-loyal subjects will snuff her out.[5]

I'm not quite sure what befell her, but I once had a hive fail utterly to replace their queen. They seemed so lost. At least they *sounded* that way. Rather than offering the unified, major-chord buzz I was used to hearing, individual bees were humming quietly to themselves, out of phase, the result a weird discombobulation. Without their queen, the workers didn't quite know what to do with themselves, and obviously, they wouldn't survive without new bees being made. To fix the problem, I rush ordered a new mated, Italian queen from a honeybee supplier in the South. Just days later, she arrived, caged along with several of her attendants, in a large, puffy envelope.

Now, introducing a new queen to a colony takes some finesse, since bees don't take well to strangers in their hive. A new queen is just that—an outsider with her own unfamiliar smell. If released from her cage too soon, this queen could be toast. Not realizing her intrinsic value to their dwindling colony, the other bees might have killed her at hello.

The queen cage—both ingenious and elegant—is designed

5 Known as "balling the queen," sometimes a small group of worker bees will crowd around their soon-to-be-former queen, moving closer and closer until she is crushed or overheats or both. Fortunately, I've never seen this happen, but other beekeepers have, noting that some workers will even bite at and tug on her antennae, wings, and legs in the process.

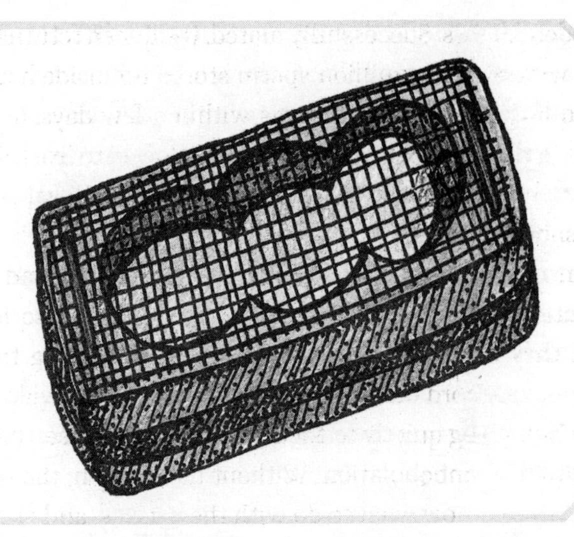

A small block of wood and some fine mesh make up this simple queen cage.

to make the introductions go more smoothly. Here's how: A queen cage is just a small block of wood with enough of its middle hollowed out to accommodate the queen and a few of her attendants. A bit of fine plastic mesh is stapled onto the side of the block with the exposed queen-holding cavity, and then two access holes—one from each end—are drilled through until they reach the hollowed-out section. The queen and her attendants are ushered into the cage, and then one of the access holes is plugged up with a small cork. The other hole is sealed with a plug made of candy, the idea being that the bees on either side of the candy plug will chew away at it until it is completely eroded. The process takes some time, and the thinking is, by the time the queen's attendants on one side of the candy plug and the thousands of bees on its other side manage to eat through the sugary stopper, the existing hive will have grown accustomed to the smell of its replacement queen. I have successfully relied on this system in the past, but it can take awhile for the bees on each side of the candy to meet.

Because this colony had been queenless for so long, I didn't have time to dally. The bees needed their sense of purpose back, I reasoned, and besides, feeling impatient and experimental, I was dying to see what would happen if I simply plopped the queen into the hive. Would my bees be so loyal to the memory of their old queen that they'd refuse to accept this newcomer? Would they kill her on sight—and as a result, waste my thirteen bucks plus shipping? I had to know.

I carefully pried out the cork and summarily dumped the queen and her attendants onto the frames in the top of the hive. What happened next astonished me, but I guess it shouldn't have. I'd read that queens sometimes "toot" or "pipe" loudly to their subjects, but I never expected to have a chance to hear it firsthand. It was a startlingly loud and clear "Whooooo-Whoooo-Who-Who-Who-Who!" As she piped, the queen pressed her midsection against the wooden tops of the honeycomb frame, serving to amplify her high-pitched, staccato calls. It sounded a bit like a kazoo being played by a boiling tea kettle. To those formerly queenless bees, I'm sure this weird vibrating whistle sounded like salvation. I'm happy to say the hive quickly got back to business as usual.

The Lazy, Yawning Drones

Sometimes I have good little talks with my next-door neighbor, Irene. She's a widow in her seventies who has halfheartedly begun to date again. The problem, though, is that there are only so many unattached, living men left to go around, and as a result, these fellows think they can take all sorts of liberties. Recently contemplating a couple of her latest rotten dates, Irene shook her head and sighed, "Men. They've got the world by the tail, haven't they?"

Yeah, pretty much. Maybe that's why I have so much contempt for the drones in my hives. Drones are male bees, whose sole purpose is to mate with any queens passing by. With that in mind, they do "work out" on occasion, strengthening their flight muscles during practice flights, but other than that, they contribute nothing. And yet they get away with *everything.* Drones don't collect pollen or nectar, but they're allowed to gorge themselves on the stuff anytime they please. They don't make beeswax or clean up around the hive either, and they're even permitted to enter other bees' hives, where they'll be wined and dined. These freeloaders don't even have stingers. In his 1901 work, *The Life of the Bee*, Maurice Maeterlinck describes them like this:

> *Indelicate and wasteful, sleek and corpulent, fully content with their idle existence as honorary lovers, they feast and carouse, throng the alleys, obstruct the passages, and hinder the work; jostling and jostled, fatuously pompous, swelled with foolish, good-natured contempt; harboring never a*

suspicion of the deep and calculating scorn wherewith the workers regard them, of the constantly growing hatred to which they give rise, or of the destiny that awaits them.

Maeterlinck continues:

> *For their pleasant slumbers they select the snuggest corners of the hive; then, rising carelessly, they flock to the open cells where the honey smells sweetest, and soil with their excrements the combs they frequent. . . . They create a prodigious stir, brush the sentry aside, overturn the cleaners, and collide with the foragers as these return laden with their humble spoil.*

Larger than worker bees but smaller than queens, drones

The large, bug-eyed bee in the center is a drone. *Rick Dietz*

are thick and muscular; their heads and eyes are enormous. They sort of look like the Nienaber twins, who, in eighth grade, were larger than the PE teacher and were already exhibiting male-pattern baldness. Despite their being quite noisy, overgrown, and kind of dumb-looking, drones do have the world by the tail. Most of the time anyway.

Cue the Barry White

A virgin queen does all of her mating within just a few days during her lifetime, and she hooks up with about a dozen drones in all. She lets them know she's oh-so-willing to mate by soaring hundreds of feet into the air and releasing pheromones that get the area males' attention—sort of like a really low-cut blouse and thong panties might in the human world. Only those drones who fly fastest and highest will be able to catch up with her, and what happens next is just bizarre.

A drone attempting to mate with the queen uses his first four legs to grab her back, and then he quickly latches on to her with his last set of legs. In order to expose his reproductive parts, he literally turns the end of his abdomen inside out, and shoves his penis into the queen's open sting chamber. It doesn't take long before a tiny explosion takes place, and it's forceful and loud enough for one to actually *hear* it. The drone shoots his sperm into the queen, after which time his genitals snap off and stay with the now-mated matriarch. The drone falls to his death, and the queen continues to mate with more drones. As if that weren't romantic enough, *The Biology of the Honey Bee* author Mark L. Winston reports, "A queen returning from

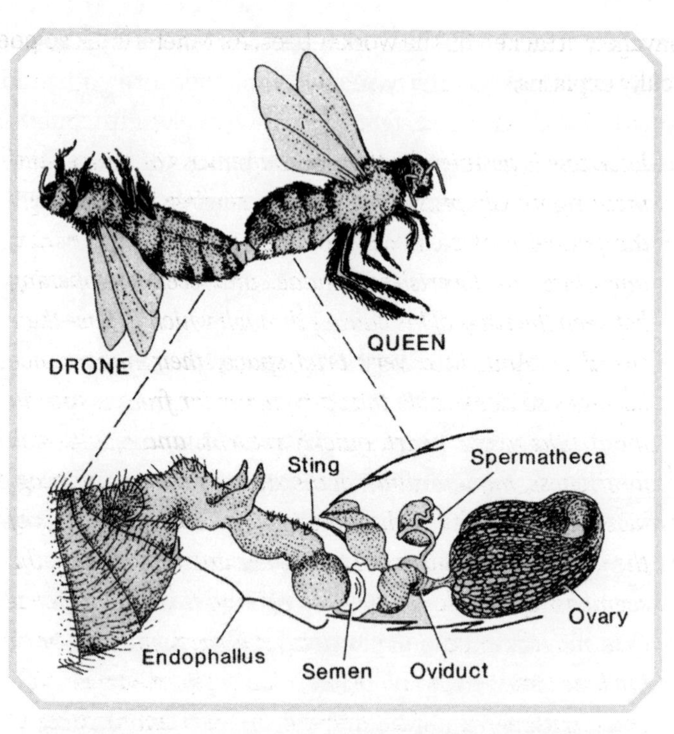

High in the sky, a drone and queen copulate. During mating, the drone's genitals will actually break off, so it's safe to say the queen gets the better end of the deal. *Harvard University Press*

a successful mating flight generally is carrying the mating sign of the last drone to mate her, and the workers which greet her lick the sign with their tongues and eventually remove it with their mandibles." Ah, sweet justice.

* * *

Drones are unlucky in at least one other way. Because they can be a significant drain on a hive trying to make it through winter, any drones still hanging around come fall are rather savagely attacked by the worker bees. As Maeterlinck so poetically explains:

> *Each one is assailed by three or four envoys of justice; and these vigorously proceed to cut off his wings, saw through the petiole that connects the abdomen with the thorax, amputate the feverish antennae, and seek an opening between the rings of his cuirass through which to pass their sword. . . . And, in a very brief space, their appearance becomes so deplorable that pity, never far from justice in the depths of our heart, quickly returns, and would seek forgiveness, though vainly, of the stern workers who recognize only nature's harsh and profound laws. The wings of the wretched creatures are torn, their antennae bitten, the segments of their legs wrenched off; and their magnificent eyes, mirrors once of the exuberant flowers, flashing back the blue light and the innocent pride of summer, now, softened by suffering, reflect only the anguish and distress of their end.*

Drones that don't immediately perish are turned out into the cold, and as each drone tries to bully his way back into the hive, worker bees alight on the drones' backs, pulling on wings and legs to make their reentry nearly impossible. All the while, the rest of the worker bees look dispassionately on as if to say, "Sorry, boys. Better luck next time." And when the drones' services are required in the spring, the queen and her workers will simply rear more of them.

Honeybee Destinies

As complex as honeybees are, it should come as no surprise that, whether it's a drone, a worker, or a queen, considerable time and effort go into the creation of each and every bee. For a long time, that creative process mystified all who tried to understand it. In Aristotle's day, some observers of the natural world believed that "bees neither copulate nor give birth to young, but that they fetch their young." In his *History of Animals*, Aristotle continues, "Some say that they fetch their young from the flower of the callyntrum; others assert that they bring them from the flower of the reed, others, from the flower of the olive." It was thought that, after happening upon the pupal bees sprinkled among the flowers, adult bees taxied them back to the hive and placed them gently into their respective cells. That's a pretty enough idea, but those ancients weren't even close.

Some historical records suggest Alexander the Great was buried in a honey-filled coffin.

The truth is, despite the marked functional differences between the three castes, honeybees all start out pretty much the same. Their lives begin inside the pearliest of eggs, which are just barely big enough to see. The queen lays these, one per cell, all by herself, and if you can spot them, they look like tiny apostrophes glistening inside their uncapped cells. Because it takes about three days before an egg will actually "hatch," any new, intact eggs a beekeeper discovers will tip him off that the queen is still alive and kicking—or at least that she was in that part of the hive very recently.

The queen can lay two kinds of eggs—fertilized and unfertilized—and which kind she lays depends on the size of the honeycomb cell into which the egg will be deposited. If she is poised over a much larger drone cell, the queen will release an unfertilized egg, which will become a drone. Otherwise, she releases eggs that have been bathed in the sperm she stored up from her previous mating flights, and those fertilized eggs usually issue worker bees.[6]

They Are What They Eat

Whether a hive's tiny bee babies are fated to be drones, workers, or even queens, nurse bees feed super-rich royal jelly to each of them at first. The viscous, milky stuff contains B vitamins, amino acids, sugars, and trace minerals, among other nutrients. As such, you'll find it in sundry health food supplements, overpriced face creams, and even soaps, but I think royal jelly does the bees much more good than it could ever do for us. (I wonder, would people still snap up those royal jelly products if they knew the magic ingredient—in part the synthesis of digested pollen and bee saliva—had been secreted from glands inside the young worker bees' heads?) Well, anyway, in his excellent book, *The Queen Must Die: And Other Affairs of Bees and Men*, author William Longgood notes, "The bees could properly say, 'You are what you eat,'" and he's absolutely right. To create a new queen bee, for instance, the nurse

6 I say "usually" because, under the right circumstances, the larva from a fertilized egg can be transformed into a new queen as needed.

bees will feed a larva copious amounts of nothing but royal jelly. But to make a worker bee or a drone? After a few days of offering only royal jelly, the nurse bees tweak their baby bee "formula," reducing the levels of protein and sugars to make what's known as "worker jelly." If a larval bee is put on a strict diet of worker jelly, she will become a worker bee. Drones, likewise, get their own specialized food, too.

Essentially, how honeybees will spend their lives—laying thousands of eggs or, perhaps, foraging for nectar—depends on the quantity and makeup of the food they're given during a critical period of their larval stage. "There's now an understanding that differences in nutrition trigger different genetic programs, and these are hormonally mediated," according

The cappings on these cells have been removed to show pupating bees in various stages during their development. *Elbert Jaycox*

to Gene E. Robinson, Swandlund Chair in the Department of Entomology at the University of Illinois at Urbana-Champaign. "We also know something about the identity of the genetic switches; insulin signaling is now known to be involved."[7]

The young nurse bees will visit each of the voracious bee larvae thousands of times, and what starts as row upon row of those tiny, glistening apostrophes gradually will grow into an army of much larger, striated, all-white grubs. The bee larvae remain in their cells, and shed their outer skeletons a number of times to allow for their continued, exponential growth.

7 Incidentally, larval bees aren't the only ones to receive the nurse bees' specialized food. Karl Crailsheim, an Austrian scientist, discovered that nurse bees also feed their older, foraging counterparts. "Foragers can't digest pollen, so their protein requirements are accounted for by getting fed from the nurse bees," Robinson adds.

This capped honeycomb contains new bees in the making.

Toward the end of the bee's larval stage, nurse bees will cap the open brood cell with wax. Sealed inside, the bee larva spins a cocoon around herself just as caterpillars do. Gradually, she begins to look more and more like a bee, complete with eyes, legs, and antennae that go from milky white to shades of honeybee black and brown. Bees spend the longest amount of their developmental periods in this pupal stage, and when it's nearly time for an adult bee to emerge from her cell, the hairs on her body fill in and her wings expand.

The last step in the creation of a honeybee is decidedly the best. A born do-it-yourselfer, the adult bee, from inside her cell, punctures her cell cap in several places as she works her way out. There have been a few times in the middle of summer when I've had one of my hives opened up, and while I was looking over a frame of capped brood and honey, I would notice an antenna or two sliding tentatively through holes in a cell's cap. The action is not unlike that of a tiny toe testing the waters of the bath or a lake. Watch long enough and you'll see a full-grown bee chew her way through and emerge, all soft and downy, from her cell.

* * *

For a worker bee, going from an egg all the way to an adult takes about 21 days. At 24 days, drones take even longer. And as for queens? About 16 days in all. Besides that silly business about the bees fetching their young from flowers, the learned thinkers of Aristotle's age got one other thing wrong: "The bee lives for six years as a rule, as an exception for seven years." Uh, not actually. A honeybee's lifespan varies depending on the time of year. Bees raised to get through the winter can hang on for up to 140 days, but the lifespan of a bee reared for summer work is just 4 to 6 weeks. Short but sweet.

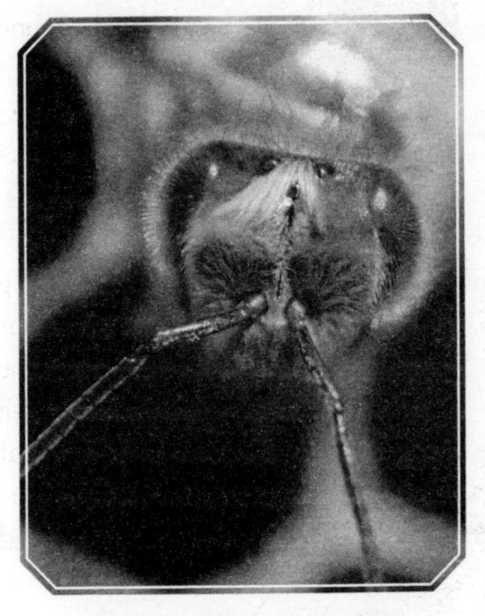

Ready to work, a new bee has chewed through the wax capping to emerge from her cell. *Rick Dietz*

CHAPTER THREE

The Bee's Knees (And Other Bee Parts)

Bees by the Numbers

Bees may not live terribly long, but they certainly do make the most of the days they have—especially the workers on nectar- and pollen-collection duty. They leave the hive about 15 times a day to get the goods, and they may clamber down the throats of as many as 100 flowers per excursion to do it. Fortunately, many hands make light work (or at least lighter work); in all, a flourishing hive can have anywhere from 60,000 to 100,000 bees.

Just how far do those bees have to fly to find what they need? That depends. Bees consigned to areas with more buildings than blossoms will fly as many as six miles on the search for nectar, but bees with access to plentiful forage may only need to fly a couple tenths of a mile. For the most part, though, bees end up flying within two or three miles of home, and at fifteen miles per hour, they make very good time.[1] Though, should you see honeybees streaming from their hives and speeding skyward, you'd think they are doing more like thirty-five miles per hour. Laden with at least half of their body weight[2] in nectar, the bees sometimes require more time for their return flights, but really, who could fault them?

* * *

Thinking like a honeybee can be an interesting exercise. Try it sometime, and you'll catch yourself scouting for potential nectar sources in your backyard, outside the post office, in city parks, or alongside the highway. You might discover that spotting every early-blooming crocus, budding silver maple, or tulip poplar doesn't come easy to you, but it's second nature to the honeybee.

When a forager finds a superior nectar source, she'll fill up, fly home, and tell the others how to locate it by doing a special dance. If the food is a couple hundred feet away, the forager performs a "round" dance, during which she runs in narrow circles, sometimes changing directions and even stopping to offer onlookers free samples of the nectar to be had. Food that is much farther away requires a more complicated "figure eight," "waggle," or "wagtail" dance, so named because its performer vigorously wags her back end at the junction between the figure eight's two rounded sections. Always using the sun

1 By the way, scientists have actually measured bees in flight to determine how many times per second a bee's wings beat, and most agree it's around 190 to 200 times per second.

2 A worker bee can weigh between 81 and 151 milligrams—about as much as several grains of rice.

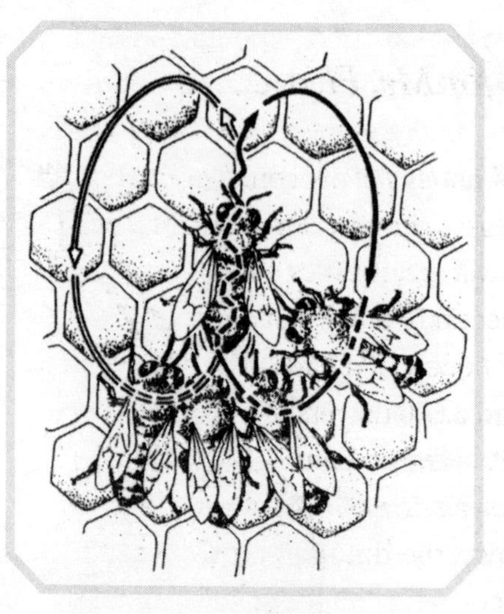

By dancing in a figure eight, a worker bee with primo information on a good nectar location tells the others how to find it. *Harvard University Press*

as her compass, a foraging bee can convey a destination's direction and distance well enough for other bees to find it.

Once reinforcements are on the scene, the hive is able to store up a lot more nectar. Besides the total number of bees collecting the stuff, the amount of nectar available is another factor that affects the total honey supply. As you know, where there are flowers, there is nectar, and whether you're a beekeeper or just an avid bee watcher, if you pay attention from year to year, you begin to see a fairly predictable ebb and flow of bloom times. Where I live, spring wildflowers give way to honeysuckle and sweet clover, which, in turn, transition to assorted fruit trees, and then fall's ironweed, goldenrod, and asters take over for those pears, peaches, and apples. While honeybees can bring in four or five pounds of the sweet syrup a day, when the nectar supply is really plentiful, they'll put up a whopping ten pounds a day as long as they have room enough to store it. Keep in mind that doesn't mean they're stockpiling ten pounds of *honey* every day. Actually, it takes four pounds of nectar—and the time and expertise of honeybees—to produce one pound of honey.

Just a Teaspoon for Me, Please . . .

Every time I spoon out a bit of honey for my coffee or oatmeal, I'm reminded that anything of real value requires hard work and a lot of it. On average, it takes about a dozen bees to gather enough nectar to make just one gloriously golden teaspoon of honey, and each of those bees must visit more than 2,600 flowers in the process. Crazier still, all those flights from the hive to the flowers and back again add up to 850 miles or so—just over the distance from New York to Chicago.

What Honey Is and Is Not

Some folks—namely, those with the quintessential seventh grade gross-out mentality—love to stir the honey pot by suggesting that the stuff is nothing more than "bee vomit," but that's an awfully oversimplified view. True, when a foraging bee slurps nectar from a flower, she does store the sugary fluid in what is known as her "honey stomach," and yes, eventually she will fly back to her hive and *regurgitate* the nectar she's collected for the other worker bees to process into honey. But wait. It's only fair to point out that the foraging bee's honey stomach isn't the same as her regular stomach. Actually, she has two stomachs, which are separated by a little valve affectionately called the honey stopper. It is this honey stopper that nicely keeps her stomachs' contents separate. Besides that, she doesn't simply pass that nectar on to the awaiting worker bees "as is." Instead, she mixes in a few enzymes, which will help to start transforming the nectar into honey, and once she's back in the hive, she shifts the contents of her honey stomach to one or more of the workers, their own honey stomachs empty and standing by.

While fresh nectar has a very high moisture content—some nectars are as much as 80 percent water—finished honey is only about 17 percent water. To get it that way, the worker bees now in charge of the load will add still more enzymes and then bring up a small amount of the nectar at a time, working it with their mouth parts as you might chew a sticky wad of gum.[3] Next, they spread the ripening honey out over many honeycomb cells and fan their wings until what was originally a fairly thin, runny liquid becomes much thicker—what we would recognize as honey. The process can take as little as one day to up to a week. As if the honey making wasn't itself miracle enough, capping the honey-filled honeycomb with beeswax is the workers' last step.

To keep their stored nectar and ripening honey from dripping out, bees create honeycomb cells, which point upward by about 13 degrees.

Waxing Philosophical

The making of beeswax brings to mind the classic chicken-and-egg conundrum—only this time it's "Which comes first,

3 Ever notice how when you blow a large bubble and leave it hanging in the air your gum seems noticeably drier? Bees expose their own "bubbles" of ripening honey to the air to help lower its moisture content.

the honey or the wax?" See, bees actually create wax by way of special glands in their abdomens, and fueling that process requires nectar that they might otherwise be using to make honey. Instead, they'll go through eight pounds of nectar to make one pound of beeswax.

The bees making the wax are workers between five and fifteen days old. To pull it off, small groups of wax producers huddle together to increase the temperature of their immediate surroundings. Not long after, they begin to secrete liquid wax along their underbellies. As the air temperature cools, the liquid wax hardens into scales, which look a bit like opaque armor plating. To put the wax to good use, each worker scrapes off her scales, pops them into her mouth, and chews them until they are soft enough to work into cell cappings or new honeycomb.

It takes approximately 450,500 of those tiny wax scales to make up a single pound of beeswax. As a nest-building material, beeswax is very sturdy stuff. Its melting point is somewhere between 143 and 148 degrees Fahrenheit, so even on the most sweltering August days, it won't liquefy (but as I can attest, hot honeycomb does disperse a most heady, sweet scent).

A little at a time, the workers will add new scales, sculpting the wax into thousands of tiny hexagons, side by side and back to back, which will shelter sundry baby bees and store loads of pollen and ripened honey. Nearly five of the teeny, six-sided cells will fit per inch, and a single pound of beeswax can be formed into 35,000 individual cells that are strong enough to hold nearly twenty-two pounds of honey. Which brings me back to that chicken-and-egg business. Certainly, it takes nectar to make wax, but it also takes *wax* to store up all that ripening honey. Hmmm . . .

Glistening wax scales form on this bee's abdomen. *Lawrence J. Connor, Wicwas Press*

This wax scale will be softened and shaped to form new honeycomb or cell cappings. *Lawrence J. Connor, Wicwas Press*

* * *

Old Man Winter's the main reason bees bother making honey at all. The typical hive needs at least sixty pounds of honey to make it through until the coming spring, but I usually leave my bees much more than that. Even as January winds whip at their hive, honeybees remain active, dipping into their stores all winter long. To generate heat, the bees stick together in a cluster, which can expand and contract depending on the temperature outside the hive.[4] Bees near the center of the cluster (and by extension, nearest to their dear queen) generate some serious heat—sometimes up to 90 degrees!—by flexing their flight muscles (but without actually flapping their wings).

Bees on the outside of the cluster serve as live insulation, but periodically, they'll make their way to the center of the cluster to warm up. As they do this, bees that previously were in the center head to the outside edges. The bees will keep trading places like this, and occasionally, they'll break into the honey stores overhead or to the sides of the cluster for a little pick-me-up. After going through all the honey in one area of the hive, the cluster of bees moves to a new section with its own honey stores. They'll keep this up until spring or until they run out of honey, whichever comes first.[5]

4 When the mercury dips many degrees below zero, overwintering bees will contract into a cluster about the size of a softball. As outside temperatures increase, the bees will allow more space between one another, so in temperatures of, say, 50 degrees and up, that same cluster can be as big as a large beach ball.

5 Or sad but true, until they *think* they've run out of honey. Sometimes bees get turned around and "forget" where their honey stores are located. I once had a colony starve to death despite the fact that they still had lots and lots of honey left. When I checked on them in the spring, I found hundreds of dead bees, their heads stuffed into exhausted honey cells, their tiny, black butts mooning me. I felt like an incredible failure.

As they are very fastidious insects, bees will break ranks on any unseasonably warm winter days to poop outside of the hive. (Unlike some people I've known, bees never mess up their own nest.) Whenever we get one of those blindingly sunny spells, the light glancing off melting ice and snow, I like to watch my bees taking what we beekeepers primly call their "cleansing flights." They fly a little ways off and then splat yellowish loads that look like yolks from the most diminutive eggs—and feel like raindrops if you are unlucky enough to be standing in the wrong place at the right time. Regardless, any time I get up close and personal with a honeybee, I'm always amazed by her elegance and complexity.

What Makes a Bee a Bee

Researchers only recently decoded the honeybee's genetic structure, and what they found surprised them. Turns out the regulation of genes in the honeybee is more similar to that of humans than it is to other insects. Considering all that bees are capable of, though, that sounds about right to me. I'm always surprised by the numbers of people who still see honeybees merely as flying hypodermics. True, they are equipped with fierce little stingers, but they also happen to have wondrous heads, thoraxes, and abdomens; the honeybee's head is for figuring things out, her thorax is mostly devoted to getting from point A to point B, and her abdomen houses assorted vital organs and, yes, that famous sting.

The Head

Technically, every bee has five eyes on its head. Two of them are gigantic compound eyes, and each of those is made up of thousands of facets. Every single facet has its own lens and sensory cells to help the bee see light, patterns, and colors. Because drones need to be able to see well enough to pick out any virgin queens flying by, they have 8,600 of these facets in each compound eye. Worker bees need to see what they're doing as well, but they don't need eyes which are quite as high-powered, so they have 6,900 facets per eye. And since the queen spends so much of her life stuck at home, laying egg after egg in the dark, she has only 3,000 to 4,000 facets in each of her compound eyes. (Incidentally, I used to envy my foraging bees. They got to work outside, alighting on springy petals,

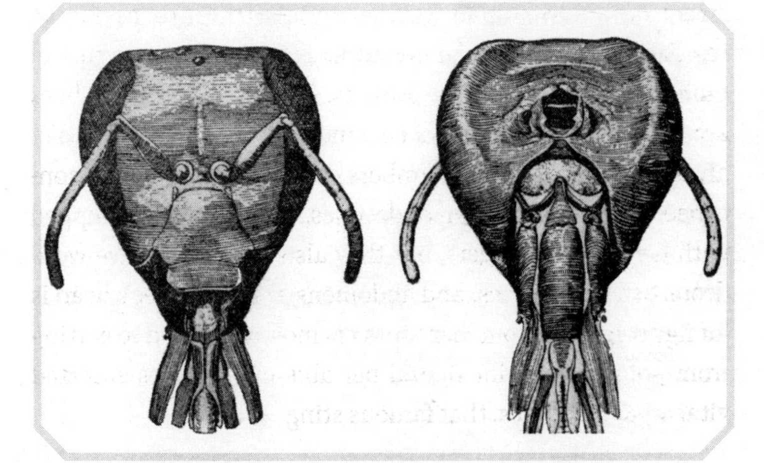

A disembodied honeybee head in great, anatomical detail.

sampling aromatic nectars, and taking in so much color. But now I know the world doesn't look quite the same to them. While bees do see in color, they don't see all of the same colors that we do. A vivid red bee balm or bright orange calendula might look dark green or black to bees, because they're blind to the red part of the color spectrum. Bees can detect the yellow, green, and blue parts of the spectrum, and they can see the ultraviolet parts, which are invisible to us.) And talk about giving someone the hairy eyeball! A honeybee's compound eyes are also full of tiny hairs, which are used to detect wind speed and direction, and so, they are handy in flight.

On average it takes about ten minutes for a foraging worker bee to amass a full load of pollen and at least three times as long to collect a full load of nectar.

A bee's other three eyes—called "ocelli"—are much more primitive. Situated in a triangle on top of the bee's head, they look like reddish or light brown dots, and they're really only useful for sensing light. Attached under the ocelli are two antennae with which the bee sniffs out nectar sources, messages from other bees, and sometimes even land mines.[6] I can say from my own experience that the honeybee's sense of smell certainly is impressive. Sometimes when I've taken a couple of frames of ripened honey for myself, I'll temporarily store them

6 Yes, land mines. In 2003, environmental chemist Jerry Bromenshenk successfully "trained" honeybees to associate the smell of explosive traces with finding food. Let loose, the bees were temporarily drawn to areas containing unexploded land mines—and unlike a curious kid or bomb-sniffing dog, they didn't get blown up in the process! In fact, in a controlled study, Bromenshenk determined that his bees could detect explosive vapors 97 percent of the time. Too bad they have their work cut out for them. Land mines take out almost 20,000 people annually—and experts say it'll take roughly 450 years to clear out all the unexploded devices still hiding out there.

in my garage—several yards away from my beehives. Despite the fact that I tightly seal the wax-covered frames in plastic bags, invariably, a handful of bees will swiftly buzz around the garage door to investigate.

In general, it's difficult to put too much over on bees, since they have large brains that are connected to several nerve centers throughout their bodies. The honeybee's head also incorporates rather menacing-looking jaws and a long, folding proboscis, which contains her tongue. From eating pollen,

Magnified under an electron microscope, the face of a worker bee looks quite furry and her antennae are clearly segmented.
Tina Carvalho, University of Hawaii

constructing honeycomb, and disposing of the dead, worker bees use their jaws for just about everything. And their soda-straw tongues are perfect for sucking the nectar out of especially hard-to-reach places as well as for feeding other bees.

The Thorax

Stretching from the head all the way to the beginning of the abdomen, an impressively long esophagus runs through the bee's thorax. This midsection is itself divided into three parts, and each of those has its own pair of legs. Also, the last two sections have one pair of wings each. One really neat thing about a bee's wings is that the front pair has a fold running along the outer edge, and the back pair has barbs that, sort of like Velcro, can hook onto that fold. This is handy when the bee wishes to rely on one large set of wings instead of her two original—and much smaller—sets.

Now to the bee's legs. Remember when *Entertainment Tonight* host Mary Hart had her gams insured for $1 million? Had she been a worker bee, that might've made a little more sense. Why? Because workers always come prepared, and every tool they need for a busy afternoon of climbing walls, looking one's best, or gathering pollen is incorporated in the design of their lovely legs.[7] (Can Ms. Hart's legs even begin to compare?) For instance, each of a worker bee's front legs includes its own brush and notched antenna cleaner. And

7 Drones and queens don't have any of the pollen- and propolis-gathering leg accessories. I doubt they'd know what to do with them if they had them anyway.

all six legs have suction cup–like pads and tiny claws for getting around and gripping things respectively. A worker's back legs come complete with tiny rakes, combs, storage compartments, and pressing tools that are perfect for collecting load after load of pollen and propolis.

The Stickiest Stuff on Earth

Besides gathering nectar, pollen, and water, bees also busy themselves with propolis or "bee glue." Bees use the gummy resin to fill any cracks and cover over foreign objects which may have gotten inside the hive.[8] Usually a butterscotch-tinted brown, propolis comes from the buds of trees and is said to have antimicrobial and antifungal properties. To collect the über-tacky stuff, a foraging worker tears bits of tree gum off with her mouth, packs those bits onto her legs, and flies her load back to the hive, where other workers will help to pull the propolis off her. The instant that load is liberated, it gets smeared somewhere within the hive. For what it's worth, I've noticed that some bees demonstrate more restraint with the propolis than others; I've had hives of bees that used it like the late Tammy Faye used mascara. In those cases, separating frames from one another and from the supers in which they're contained requires patience and a lot of prying with one's steel hive tool.

8 I once propped a small stick under my hive's top cover to improve the air circulation during a particularly hot spell. At some point, though, the stick slipped out of place. The next time I checked the hive, I noticed the bees had rendered the stick—now covered with an impossibly thick layer of propolis—nearly unrecognizable. I have heard that bees will propolize anything that doesn't belong, including large beetles and even the errant mouse.

The Abdomen

The largest of the honeybee's main parts, the abdomen is so jam-packed with organs that I'm not sure just how they all manage to fit. The heart, for one, is a *looooong* tube running through the entire abdomen and through the thorax to the head, and it loosely circulates haemolymph—the bee equivalent of blood—throughout the bee's body cavity instead of through dedicated blood vessels. Carrying on alongside the heart is that internal briefcase, the honey stomach. It's connected to the honeybee's true stomach in the part of the abdomen that's closest to the thorax.[9] A series of twisting tubes comes next, transferring waste from the bee's blood to the intestine, on to the rectum, and out the anus (and often, all over my bee suit).

Breathing is another bee bodily function that you can actually see for yourself. The outsides of a bee's abdomen and thorax have tiny holes, called "spiracles," which lead from the great outdoors to air sacs and trachea inside the bee. Air can pass freely into the bee while she is still, but by rhythmically pumping her abdomen, a bee can force extra air into her body if she needs to.[10]

A honeybee's abdomen also includes his or her sex bits. Worker bees have small ovaries and undeveloped sperm stor-

9 Designed to hold enough fuel for long flights and overwintering, the fairly long stomach, or ventriculus, is coiled up inside the abdomen. If you could straighten it out, it would be about twice as long as the bee herself.

10 Unfortunately, harmful chemical agents like pesticides and herbicides that may be drifting through the air can also pass into a bee's body, so to stay on the safe side, most beekeepers simply don't use them.

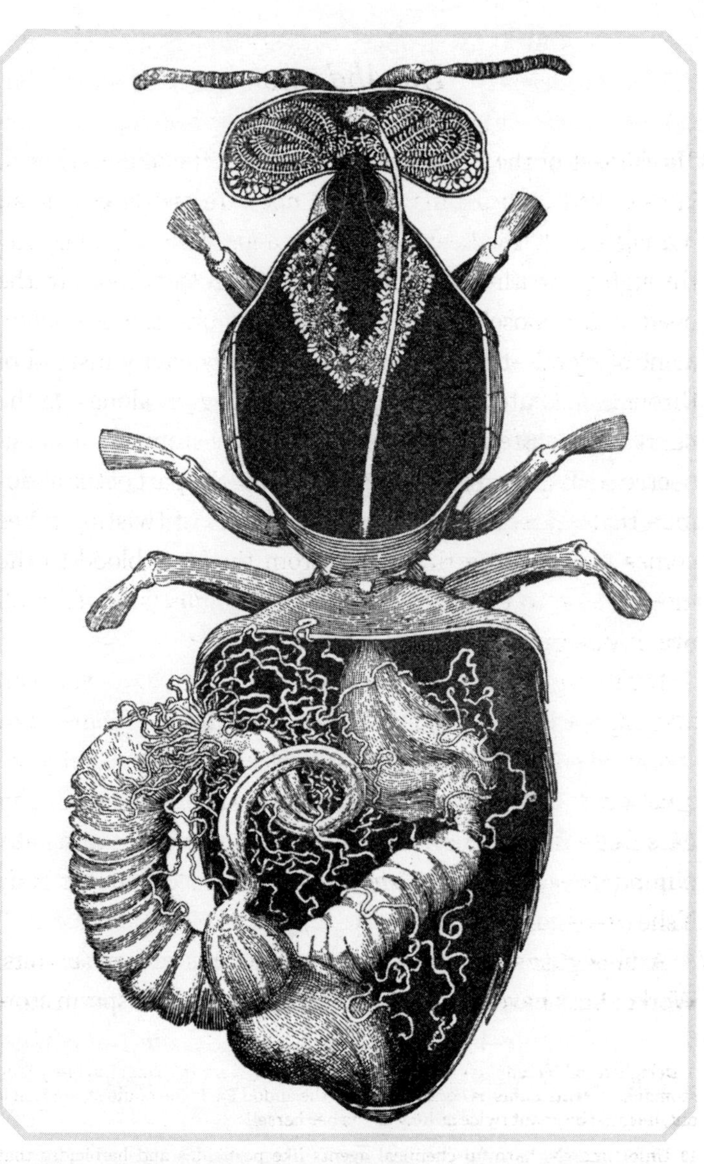

The honeybee's salivary glands, honey stomach, intestines, and other digestive organs are on display in this detailed woodcut.

age areas, while those of queen bees are understandably large. And as for drones? They have penises and testes, which they manage to keep tucked away until it's time to do their part for the good of future colonies.

Gland Central Station

The abdomen's also the spot for a number of the worker bee's specialized glands, including the wax glands; the scent glands, which produce the pheromones honeybees use to communicate with one another; and like it or not, the poison gland, which supplies the venom for the honeybee's sting.[11]

The Sting

Until I was five or six, I had been perennially barefoot in summer. And so it would be just a matter of time before I'd experience my first bee sting. A baby food jar with a half inch of hose water in one hand, my mashing stick in the other, I was running over to a stand of four o'clocks to make "perfume" out of their open flowers. Along the way I must have stepped on a worker bee as she foraged among the ample dandelions and clover in our yard. In an instant I'd dropped to the ground and inspected my foot: pulsating unpleasantly, the tiny stinger drove itself deeper into the soft pad next to my toes. I noticed

11 A bee may send out a scent message to alert her fellow workers to a particularly good water source or, perhaps, to corral others during a swarm.

bees everywhere after that, and it would be decades before I'd again traipse shoeless on a lawn.

Such is the power of the honeybee's sting. Fortunately, I'm not one of the unlucky few who are allergic to bee venom, and anyway, by now I have a better sense of when and why bees will sting.[12] First off, foraging and swarming bees really aren't so apt to sting unless you go out of your way to bother them, and as they completely lack stingers, drones can't sting at all. But if a colony's guard bees decide that the hive is being threatened in some way, they will most certainly sting. That threat might come from skunks skulking around the hive for a midnight snack, or perhaps it's an inexperienced beekeeper, handling them with all the finesse and dexterity of a sea cow.[13] Some bees are just more mellow than others. Italians, with fantastic, pale gold banding, were the first honeybees I ever worked with, and each time I popped open their hive and gazed inside, they barely seemed to notice me. About

Applied directly to the wound, a tincture of opium was one common bee sting remedy in England during the 1800s.

12 After I've been stung, I *do* feel achy and shaken, as if I've been in a car accident, but people with very strong allergic reactions to bee venom fare much worse, sometimes going into shock, having difficulty breathing, or even suffocating. About 1 out of every 1,000 people is allergic to bee stings.

13 When I started keeping bees, I originally placed my hives directly on the ground, inadvertently subjecting my girls to insect-eating skunks, which raid colonies by scratching at the hive entrance until guard bees come out to investigate—only to be smashed and eaten. The skunk then scratches some more, drawing out additional curious bees who meet with the same fate. And on and on it goes. Sure, the bees can and do sting the skunk as it dines, but they can't inflict much damage through the thick fur on the animal's head and back. Thus, a poorly positioned beehive becomes a veritable slot machine that always pays out. But fixing the problem's a snap. In my own apiary, I elevated the hives just enough to make it a fair fight. Now when the skunks come around, they must stand on their hind legs to reach the hive entrance, and the bees are able to get some good jabs in on the skunks' delicate undersides.

the time they'd lulled me into complacency, though, everything changed. The guard bees who once tolerated me now hurled themselves ceaselessly at my veil, and I was stung at least once or twice every time I inspected the hive.

Coming to fear my new pastime, I described the problem to one of my beekeeping mentors, who told me my bees had turned "uppity." In short, the once-docile bees had likely replaced the Italian queen I'd installed that spring with someone new—a virgin queen who subsequently mated with a hodgepodge of feral drones in the area, producing distinctly more aggressive bees.[14] To fix the problem, I would have to mail order another mated Italian queen, so that I could replace their replacement. To do that, I would have to find the old queen and kill her, so that the new queen could reign supreme.[15] Then, as thousands of my new queen's offspring began to emerge, the tenor of the hive would again be good-natured. But the idea of killing an otherwise perfectly good queen never sat well with me. Instead, I decided I would work with my bees as they were and take extra precautions—like duct taping the cuffs of my pants and my shirtsleeves to keep those hypervigilant guard bees out.

Bees are also more likely to rally and sting once they get wind of the alarm pheromones released when someone

14 Bees will begin raising several new queens at once if their current queen meets some untimely end or if they are generally underwhelmed by her performance. Bees from very large, successful hives will also raise a new queen if part of the colony intends to split off and set up somewhere else.

15 Rather than admit that they themselves had much to do with an old queen's death, some beekeepers will shrug their shoulders and say this most despicable thing or something close to it: "Well, I gave her the ol' hive tool test, and she failed," which actually means the beekeeper found the old queen and squished her with the top of his steel hive tool. Some test.

else's stinger is deployed, so whenever I am stung, I direct a few puffs from the smoker onto the stung area to help mask that "Attack! Attack!" message. I remember one day when I was working in the hive by myself, my smoker went out, and the bees were trying to tell me I had overstayed my welcome. One stung the arm of my shirt and then another and another rammed the same spot. I took off the shirt, tossed it away from me onto the ground, and the bees stayed with it, reacting strongly to the alarm pheromones that had been released.

If you've ever been stung, you know alarm pheromone isn't the only thing that comes with a bee's stinger. Produced in the worker bee's poison gland and then transferred to her poison sac, bee venom is the real culprit behind the acute pain followed by the soreness and swelling that characterize a honeybee sting. The stinger itself is V-shaped and barbed like a fishhook, so when it punctures thick skin, it gets stuck.[16] As the bee tries to get away, her stinger and poison sac stay behind. Missing her back end is inconvenient to say the least, and soon the worker bee will die.

This is a bee sting.
Tina Carvalho, University of Hawaii

16 When a bee stings other insects, though, she's able to retain her sting and use it multiple times. That's because it's much easier to free one's stinger from brittle exoskeleton than from a mammal's tough skin.

* * *

This is Sting. *Getty Images*

Before he became the Police's front man, Sting was just a guy named Gordon Matthew Sumner who sometimes wore a black-and-yellow-striped sweater, which made him look like a bee. As a result, a fellow musician took to calling him "Sting," and the name stuck.

Dang It!

You've been stung. So now what? Cursing loudly seems to help me, but first things first. If there are other bees near enough to be provoked by the alarm pheromones, go indoors right away. Because the poison sac attached to the stinger will continue to pump venom into the sting site even *after* the bee responsible for the deed has crawled off somewhere to die, the sooner the stinger is out of you, the better. To avoid forcing extra venom into your body, don't pull the stinger straight out. Instead, use your fingernail or the cover of a matchbook to scrape it out.

Apply an ice pack to help ease pain and swelling, and take

heart: A little bee venom may not be all bad. Devotees of bee venom therapy go out of their way to be stung by honeybees for the treatment of rheumatoid arthritis, multiple sclerosis, and myriad other illnesses, but scientists have yet to establish the efficacy of bee venom treatment.

Besides that, it takes a lot of bee venom to do most of us in—general consensus is between 8 and 10 bee stings per pound of body weight—and one fellow was stung 2,443 times and lived to tell the tale.

CHAPTER FOUR

Bees and the Bigger Picture

Bee Souls, Legends, and Lore

In my earliest days as a burgeoning beekeeper, I learned this quaint rhyme:

> *A swarm of bees in May,*
> *Is worth a load of hay;*
> *A swarm of bees in June,*
> *Is worth a silver spoon;*
> *A swarm in July,*
> *Is not worth a fly.*

In fact, each time I mentioned that I'd begun keeping bees, someone invariably would recite those lines to me. A few were fellow beekeepers. Others were septuagenarians with departed fathers or grandfathers who had once tended apiaries and passed the saying on. Still others just seemed to have been born with that little rhyme stuck in their heads. Funny, it seemed none of them could recite the last part without smiling wistfully at the end.

Turns out, the rhyme had knocked about France, Germany, and the United Kingdom long before it traveled to early America, and naturally, it changed a bit along the way. In some regions, a swarm of bees in May might be worth a single sheep or a particularly bounteous milk cow, for instance. And other folks chose to end with *A swarm in July? Let them fly* instead.

Anyway, when I originally heard that rhyme, I admit I didn't quite understand just what it meant. I mean, a swarm of bees is a swarm of bees, right? Well, not quite. Essentially, the earlier one finds a swarm, the better off he'll be, since bees need enough time to build up their honey stores and winter bee populations. Find a swarm in May, and the worker bees have all summer to gather nectar, and the queen has time enough to make herself comfortable in her new home. Find them much later, and they'll never come up with enough honey and bee reinforcements to last through the winter.

Even so, I have a very hard time letting those ill-timed July swarms fly off to certain ruin, so I altered that old rhyme's ending: *A swarm in July? I'll give them a try!* And do I ever try. I mix up pan after pan of sugar syrup on my kitchen stove and crunch through the snow to feed the bees regularly until late spring.[1] (Ever tried to keep a match lit on a windy day? Keeping

1 Although the idea is strangely appealing, I do not feed my overwintering bees with a teeny, tiny spoon. Instead, I pour the syrup into a hive body feeder, which fits inside an empty super. The shallow reservoir has two honeybee "walkways" topped with plastic grates, so that the bees can safely have a nip of nectar without the risk of drowning in too much of a good thing.

a weak cluster of bees going through winter is a lot like that.) Often, I'll kneel down and press my ear against the hives' cold exteriors to listen for my bees' reassuring hum. And if I can't quite hear them, I'll knock-knock-knock along the sides of my supers to get their attention.[2] About half the time, the late swarm survives, and I feel relieved and proud. There is, too, the strongest sense that I've quietly pulled off something big.

And who knows, maybe I have. Depending on whom you ask, those wayward bees weren't *just* bees, after all. Countless

2 "Super" is just a fancy term for one of the boxy sections containing the removable frames of honeycomb. A hive can be made up of one or two supers or as many as five or more depending on the colony's need for brood space and honey storage.

ancient civilizations believed that dead men's souls went on to inhabit the bodies of honeybees, and people in some parts of India still think that's true. Perhaps I nourished and sheltered the ethereal essences of Amelia Earhart, Mahatma Gandhi, or, say, Mae West. At the height of summer, sometimes my hives are up to 50,000 "souls" strong.

If you believe in souls in the first place, that is. I am your garden-variety agnostic—something that had clearly troubled Dr. P., one of my beekeeping mentors. A retired doctor and a devout Christian, he reverently took me through his own beekeeping rituals, showing me how to light the bee smoker and keep it lit and how to inspect a hive without disturbing its occupants too much.[3] One summer afternoon under cotton candy clouds, he'd used his hive tool to loosen the propolis the worker bees quite generously slathered between the hive cover and the edge of the top super, so that we could look in on the colony. It was one of the first times I'd had the opportunity to regard honeybees from that vantage point. I watched workers just back from nectar runs scurry through, their honey sacs brimming. Others unpacked rather cumbersome-looking pollen loads from the baskets on their

3 Most beekeeping smokers look alike—a bellows is attached to a metal can, and the can portion is topped with a hinged, funnel-shaped lid. To make smoke, I'll light some crumpled newspaper or dried leaves and place them in the bottom of the can. Next, I place a few hunks of old wood over the flames and gently pump the bellows. Once the wood bits cook down to glowing embers, I close the fitted lid. If I've done it right, a torrent of smoke should roll out of the tip of the smoker, and it only takes a little to help calm my bees, direct their movements in the hive, and temporarily disrupt their ability to communicate with one another.

legs. There were bees building new sections of honeycomb, drones, characteristically, doing not much of anything, and just a few guard bees showing some interest in us. About the time I decided the hive was its own buzzing, golden city, Dr. P. remarked, "You just can't look at this and doubt that there is a divine Creator at work here." I'm pretty sure he thought those honeybees could save my soul.

That's a tall order, but bees have been imbued with serious power over the centuries. Their handiwork shows up several times in the Bible with its references to a land flowing with milk and honey, a land of olive trees and honey, streams of honey and butter, and so on. And the Koran promises, among other things, a paradise complete with "rivers of honey clarified" for those who guard against evil. It's no wonder, really, as honeybees have brought sweetness and light to humanity for as long as anyone can remember. In *The Sacred Bee in Ancient Times and Folklore*, Hilda M. Ransome recounts a Russian fable that drives this point home:

> *When God had created all the animals, he gathered them together in order to bless them and to give them rules for their lives. The bee came also, tired and wounded by her long journey. As a reward for her obedience God ordained that her blood and sweat should become honey, and that her wax should burn in the churches.*

And, Ransome adds, a legend from France's Brittany region "tells that bees were created from the tears which Jesus shed on the cross. Not one fell on the ground, but they all became these winged creatures, which flew away with the Saviour's blessing to take sweetness to men." Bees were also said to have flown out of Christ's belly button while he was suffering on the cross. Other fables suggest honeybees emerged from a hole in the middle of his forehead, or that they came from his blood. For their part, the Greeks thought bees naturally emanated from dead oxen.

Those last few are a little creepy, if you ask me, but bees do have a dark side. Case in point, some of the first voodoo dolls were made from beeswax. Make deep cuts in the soft, yellow figures, and you'd cause great hardship and injury to your intended targets. Many voodoo practitioners also believed honey—when smeared on a person from head to heel—could keep ghosts away. What's more, they maintained, if you were to form small cakes of honey, amaranth seed, and a dash of whiskey, and then eat the cakes just before the new moon, you would be able to see into the future.[4]

Understandably, the ability to produce such indispensable honey and beeswax went straight to their heads, and as one old folktale has it, the bees asked God to give them fancy, silver hives and the capacity to sting men to death. Needless to say, he didn't think much of their collective attitude. Turning the

4 I could not resist trying this for myself, so I ordered amaranth from a specialty health food store, bought some Jameson Irish whiskey, and dipped out some honey from my own stash. (Should you wish to try this at home, you must first "pop" the amaranth like popcorn on the stove. Keep the frying pan moving and top with a lid to keep the popped amaranth from flying out.) I mixed the now-popped amaranth with just enough honey and whiskey to make a paste. I suppose I could've stuck this in the oven to harden into little cakes, but the future could slip right by while I waited. So, I spooned the mixture into my mouth under a waning June moon and waited to see what was in store for me. And I did see, sort of. I saw myself going back into the house, disappointed and wondering what to do with the sticky mess I made. Maybe I should have baked it after all, or maybe I failed to adequately stir the stuff. Maybe your results will be better?

tables on them, God housed them in uninspired straw hives and caused each sting to be fatal not to man but, instead, to the bee wielding it. Ouch.

You'd think that would cause them to use their stingers more judiciously than they do, but it doesn't seem to have helped—at least not in my own apiary. As it happens, I've been stung scores of times right through my protective jumpsuit and gloves, and every time I wonder how it is that something so tiny can create such an inferno under the skin. Looking back, I now know that I had those stings coming, as there are certain things you just don't do. Like standing directly in front of a hive, smack in the middle of the bees' flight path. Or trying to inspect the hives on rainy or overcast days when, instead of zooming from bloom to bloom, the crotchety bees are stuck indoors. (Try dodging raindrops as big as you are, and you begin to understand why even the *threat* of rain is enough to put nectar and pollen gathering on hold.)

In general, bees sting only to protect the hive, and they're quite agreeable as long as they're handled with dignity. Good beekeepers stay calm and move at what Dr. P. likes to call "bee speed"—that's slow and steady, almost as if you're underwater. (I'd previously moved like an epileptic at a rave.) Now maybe he was simply projecting his own biases, but another old apiarist told me that bees also strenuously disapprove of adultery, foul language, and drinking to excess, and that they make their convictions known with the sharp thrusts of their stingers. Supposedly, bees express their disapproval, too, if they aren't told at once of any new births, weddings, or deaths in the family. Beekeepers in France, Germany, the United Kingdom, and early America believed that unless one's honeybees were kept in the loop, perfectly healthy hives would in short order deteriorate until they had completely collapsed. To stave such trouble off, hives that had recently lost their keepers were quietly reasoned with. To ease the transition, their new caretakers might even reposition the hives slightly and drape them with black crepe. Such is the bond between beekeeper and bees.

The Letters Bee

Take in the first act of William Shakespeare's *King Henry V*, and you realize the Bard of Avon had a fairly good sense of the lives of honeybees. All right, so his Archbishop of Canterbury doesn't realize the honeybee king is really a *queen* and that nearly all her subjects are *female*. At least he does get the part of the drone just right . . .

They have a king and officers of sorts;
Where some, like magistrates, correct at home,
Others, like merchants, venture trade abroad,
Others, like soldiers, armed in their stings,
Make boot upon the summer's velvet buds,
Which pillage they with merry march bring home
To the tent-royal of their emperor;
Who, busied in his majesty, surveys
The singing masons building roofs of gold,
The civil citizens kneading up the honey,
The poor mechanic porters crowding in
Their heavy burdens at his narrow gate,
The sad-eyed justice, with his surly hum,

Delivering o'er to executors pale
The lazy yawning drone.

In our most sublime literature, the bee materializes again and again. She and her hymenopteran sisters buzz amicably through Walt Whitman's "A July Afternoon by the Pond":

The fervent heat, but so much more endurable in this pure air—the white and pink pond-blossoms, with great heart-shaped leaves; the glassy waters of the creek, the banks, with dense bushery, and the picturesque beeches and shade and turf; the tremulous, reedy call of some bird from recesses, breaking the warm, indolent, half-voluptuous silence; an occasional wasp, hornet, honey-bee or bumble (they hover near my hands or face, yet annoy me not, nor I them, as they appear to examine, find nothing, and away they go)—

And honeybees are William Butler Yeats's invited guests at "The Lake Isle of Innisfree":

I will arise and go now, and go to Innisfree,
And a small cabin build there, of clay and wattles made;
Nine bean rows will I have there, a hive for the honey bee,
And live alone in the bee-loud glade.

The honeybee has even made her way into our simplest figures of speech. Consider, bees don't exactly have knees, but serving to denote the best of the best, the "bee's knees" has long been in our lexicon. So, too, has "a bee in his bonnet," which usually describes a person's peculiar preoccupation with something or other. And who hasn't made a "beeline" for the dessert table or, perhaps, the nearest exit? That illustrative phrase describes the ever-efficient honeybee's tendency to fly the straightest, most direct route to good sources of nectar and pollen and then back home again. And the old "busy as a bee" needs no explanation.

Bee Movies, Bee Beards, and Bee-kinis

Somewhere during my 1970s girlhood, honeybees got slapped with a wretched reputation. Switch on the TV back then, and you faced the menacing, minor-key buzz of sundry B-movie "killer" bees. I was only six or seven when I saw one of the most skin-crawling, made-for-television flicks ever, *The Savage Bees*, complete with the tagline, "They're coming this way, not to make honey, but to kill!" I dimly remember a swarm of Africanized bees had made its way from a foreign freighter to Mardi Gras, and chaos ensued. Onslaught after onslaught of live honeybees darkened the sky, and still more bounced off a shiny red VW bug, terrorizing the pretty lady trapped inside. I couldn't look away.

Made long before computer-generated animation was de rigueur, *The Savage Bees* employed thousands of real bees and their handler, Norm Gary, to induce apian dread. Several more films would come along, including *The Savage Bees* sequel—*Terror Out of the Sky*, along with *The Swarm*, *Killer Bees*, and *Invasion of the Bee Girls*, among others. A research scientist and professional "bee wrangler" for nearly forty years, Gary's worked on most all of them—plus more mainstream offerings

such as *Fried Green Tomatoes*, *My Girl*, *Man of the House*, *The X-Files*, and *Candyman* and many television shows including *The Tonight Show*, *That's Incredible*, and *Jackass*. (By strategically placing a few dots of queen bee pheromone on a nearly naked Johnny Knoxville, Gary lured some 15,000 honeybees to the *Jackass* star's crotch to create a "bee-kini.")

In all, Gary estimates he's been stung over 75,000 times, and one can't help but wonder: What is it about bees that makes grown men want to monkey around with them for the amusement of others? With the threat of multiple, venomous stings, it must surely be the element of danger involved. Or perhaps, with all that humming like some insectile Gregorian chant and the occasional, curling puff of smoke, it's the honeybee's mystique that draws audiences in. Whatever the reasons, entertaining with insects has been going strong for a couple hundred years.[5] As early as the 1770s, for

5 "Entertaining Entomology: Insects and Insect Performers in the Eighteenth Century," by Deirdre Coleman, University of Sydney, from *Eighteenth-Century Life*, volume 30, number 3, Summer 2006.

instance, another apiarist and master showman, Daniel Wildman, amazed Londoners with his own honeybee spectaculars. According to Richard D. Altick's *Shows of London*, Wildman was able to juggle multiple swarms of bees and, "When he fired a pistol, the bees would break ranks, half of them marching over a table, and the others returning to the hive." With one foot in the stirrups and the other on the animal's neck, the unusual performer even rode a horse while wearing a great mask of live bees on his face.

By the 1940s and 1950s, there was a spate of more "traditional" showmen who seemed to prefer the plain old bee beard to the fancy bee masks or bee-kinis. Just don't ask why. "For apparently no better reason other than he feels like it, Apiarist Ethan Andrew of Govans, Md. will kill a few hours every once in a while by covering his head and neck with bees. The value of this bee headgear seems to be largely aesthetic, inasmuch as it is neither warm nor waterproof. Also, it is difficult to put on," a 1946 issue of *Life Magazine* reports.

Difficult, yes, but not impossible. In Andrew's case, he calmed his bees by sprinkling a little water on them, and then he positioned himself close enough to his supers that the bees would climb onto his chin. Interestingly, "The rate at which they crawl up on him depends on how much he is sweating. Andrew has discovered that bees do not like to walk in human perspiration." But that doesn't mean that they won't.

The secret behind every good bee beard? Careful manipulation of the queen. On very sticky days, Andrew would pluck up the queen, stick her in a tiny cage, and hold the cage between his teeth. The technique "usually does the trick because Andrew has also discovered that bees, in order to be near their queen, will wade through sweat up to their hips." Truly, wherever the queen is, that's exactly where the rest of the hive wishes to be.[6]

Easily entertained, Ethan Andrew regularly invited his honeybees to cling to his face. *Getty Images*

But those were simpler times. It takes a lot more than a few thousand bees clinging to a man's head to impress today's sophisticated audiences. One fellow positioned thirty separately caged queens on his person to entice over 300,000 bees—totaling 87.5 pounds—to land on him at one time, setting a Guinness World Record.[7]

Norm Gary, likewise, pursued his own world record: "I had trained bees to do everything else—to light on someone's nose, to crawl under their clothing, to cluster all over them.

6 In part that's because queen bees secrete "queen substance," a pheromone that keeps workers calm and unified. Weakened or aging queens release less and less of the stuff, and diminished amounts of queen substance in the hive will trigger the workers to begin rearing a new queen.

7 Mark Biancaniello accomplished the feat in July 1998.

After a career of doing stunts with bees, I was searching for something that was absolutely unique," he recalls. He certainly found it. Gary set the Guinness World Record for training 109 honeybees to fly into his mouth.[8]

To make it count, he couldn't just gobble a handful of bees. Instead, he would build on his years of research, attracting bees to a specific destination where they were rewarded with a fragrant sugar syrup. Eventually, he put himself in place of the bee feeding station to which they'd become accustomed: "I just sat in a chair, leaned back, and opened my mouth. I had this fragrance of the food in my mouth, and I exhaled, and of course, they were in a feeding frenzy. They just zoomed right into my mouth." In just a few seconds, Gary's mouth was so full that he had to start closing his lips to stop the honeybee ingress. "That was a very delicate task with bees crawling all over my mouth and tickling. They really tickle! So, I finally got my mouth closed. I tried to smile, but I couldn't."

Gary wasn't stung during the stunt, and the record-setting bees didn't seem to notice or care about their accomplishment. For better or worse, honeybees are often much too busy to be bothered with personal reflection.

8 As if that weren't enough, Gary actually closed his mouth, temporarily trapping all those bees, and kept it shut for ten full seconds. One-Mississippi, two-Mississippi, three-Mississippi . . . When I asked him how he kept them from crawling down his throat, this is just what he told me: "Let's say you're drinking from a canteen, and you get five people who want to drink from it. You don't want to get any germs, so you sort of tilt your head back, open your mouth, and dump a bit of water, and you close the back of your throat in a way so that you don't choke. That's what you do. Mentally, you close the back of the throat."

CHAPTER FIVE

Honeybee Perils

Why Bees Matter

Sit just to one side of a beehive when pollen is plentiful, and you'll see wave after wave of honeybees returning, their leg baskets brimming with arrestingly bright, tidily packed pollen loads. Maneuvering with the extra weight must be difficult; most bees glide gracefully inside the hive, but occasionally, an ungainly one will misjudge the landing and tumble into the coming-and-going others. The color ranges and amounts of pollen that each bee is able to manage are a marvel. There are the creamy whites and smiley face yellows, sure, but there are delicate grays, blues, purples, and blacks as well. Bees gather these colorful loads one minuscule grain at a time, and as the bees work, excess pollen grains get stuck to the honeybees' tiny hairs, which, in turn, help transfer the pollen from male flower parts to female ones, making honeybees inadvertently instrumental in a plant's production of seeds and fruit.

It happens that bees are responsible for pollinating 80 percent of all the plants on the planet. From almonds and apples to blueberries and coffee, that includes about 90 different food crops that we depend on. It's widely believed that one out of every three or four bites of food we eat is the result of the activity of honeybees. Without their pollination help, our diets would be a lot less colorful. Instead of loads of fire-engine red peppers, neon orange pumpkins, and great, green avocados, cucumbers, kiwis, or broccoli, for instance, we'd be stuck eating more plain Jane grains, rice, and fish than we could stomach.[1]

Of course, hived honeybee colonies aren't the only pollinators around. Birds, bats, and the wind get some credit, along with feral bees like bumblebees, sweat bees, and carpenter bees. There was a time when wild bees were the preeminent agricultural pollinators, and farmers could take natural pollination for granted. Ironically, though, as increasing amounts of forested land—where many feral bees prefer to live—was turned into farmland, farmers had to start renting beekeepers' managed hives to pollinate their crops. That wasn't such a big deal when bees and beekeepers were plentiful. There were nearly five million honeybee colonies in the United States after World War II. But in the late 1950s and 1960s that number began to plummet, and these days, it's estimated that there are just under two and a half million managed honeybee colonies in the United States. In part, modern agriculture's reliance on chemical pesticides is to blame, but honeybees face still other challenges.

1 Even ice cream would be affected, according to the makers of Häagen-Dazs. The company claims bees are responsible for 40 percent of its flavors, including strawberry, toasted pecan, and banana split.

Pests and Pestilence

Every beekeeper has a run-in with wax moths at least once, and wax moth larvae can really wreck a hive. Eating their way through beeswax comb, the cream-colored larvae leave twisting tunnels in place of the bees' orderly hexagonal cells, and when they're ready, they spin webby cocoons in which they'll pupate until it's time to emerge as whitish-tan moths. Usually, if a hive has a strong population of bees, the bees can keep wax moths from doing such damage.

But for any empty honeycomb in storage, it's another story entirely. The number of supers on one's hives is typically reduced for the winter, and the extra supers and the frames they contain must be tucked away until spring. Unless those frames are stuck in a large freezer or kept in an area which consistently stays under 40 degrees or so, the wax moths, now left unchecked, will find a way to turn once pristine honeycomb into crumbly, brown nothingness. As a result, when it's time to build up the size of the hive in spring, one's honeybees must spend valuable time and energy rebuilding the wax honeycomb on any wax moth–damaged frames, instead of storing as much nectar as they can to ripen into honey.

The small hive beetle is another, more problematic, adversary. Originating in Africa, the six-legged insect is a third of the size of the average worker bee. Small hive beetles are dark-colored, with hard, shiny backs that stingers can't seem to penetrate, and although bees are able to recognize them as intruders, the hive beetles can get the better of them in still another way. In their defense, the honeybees do cordon off groups of adult beetles, effectively "circling the wagons" to prevent the beetles from running amok. But the beetles don't seem to mind too much; when they get hungry, they simply rub the mouthparts of the bees guarding them. This stimulates the bees' natural feeding response, resulting in endless free meals.[2] Hive beetles that are able to get far enough away from the guard bees to lay eggs, do so on the honeycomb cells, and any emerging hive beetle larvae feed on the honeycomb itself, along with honey, pollen, eggs, and larval bees. As with wax moths, keeping one's hives as strong as possible is the best defense against the small hive beetle, but many beekeepers have found the need to use chemical controls.

Small but Mite-y

They may not have Bela Lugosi's screen presence, but vampires, unfortunately, do loom large in the honeybee world. As far as insect pests go, the Varroa or "vampire" mite is easily the worst of the worst. Eight-legged and golden-brown, a single Varroa destructor is only about the size of a worker bee's eye, but these tiny, external parasites nearly destroyed the beekeeping industry in the 1980s and 1990s. Despite their best efforts to contain the problem with chemical miticide treatments, commercial beekeepers typically lost about half of their colonies to the mites each year during the height of the epidemic, and Varroa mites still do significant damage today. When they aren't busy laying their eggs inside honey-

2 Bees are happy to share with one another at any given time, so one bee might touch her antennae to the mouthparts of another to indicate that she'd like some of that bee's food.

bee brood cells, female Varroa mites attach themselves to the backs of adult honeybees and slowly suck out the bees' blood. And as they grow inside honeybee brood cells, the young Varroa mites feed on the larval bees contained therein. Each new generation of Varroa mites does likewise, and before long, their populations have exploded to the point that even a strong honeybee colony can't survive.

* * *

There's nothing better than personal experience to make the abstract real. I'd heard the devastating statistics, and I even knew beekeepers who had sustained losses, but foolishly, I never thought the Varroa mite would actually hit my apiary. The day I discovered I'd lost one of my own hives to Varroa mites was as sad as it was strange. I'd suited up, got the smoker lit, and popped the top off the hive expecting to see thousands of my girls buzzing at their business as usual. Instead it was eerily quiet in the hive. I set section after empty section aside to get to the brood areas, where I found just a couple hundred bees left. On my hands and knees for a better look, I saw the awful mites taking their free ride on the backs of my beautiful bees. It was much too late to do anything for them. By the next week, the hive was dead, and I had one big mess on my hands.

One other noteworthy parasite is the tracheal mite. Unlike Varroa, tracheal mites actually infest the airways of honeybees. These mites are no bigger than the tiniest speck of dust; so to see them, one must dissect an unfortunate bee or two to examine under a microscope.

Some honeybee afflictions are slightly easier to spot, but considering their severity, that's cold comfort indeed. American

Foulbrood, or AFB, is a highly contagious disease caused by the *Bacillus larvae* bacterium, which produces spores that are deadly to bees still in the early brood stage. AFB causes larval bees to deliquesce, and the effect is as disgusting as you might expect. Worse yet, because AFB spores can hang around for forty years or more, beekeepers with infected hives must kill their bees, burn their brood frames, bees, and wax, and bury the ashes.

European Foulbrood is caused by the *Streptococcus pluton* bacterium, and it also affects very young, larval bees, but unlike the American Foulbrood–causing bacterium, *Streptococcus pluton* doesn't form spores. That means beehives hit with European Foulbrood don't have to be destroyed. Larvae with EFB often look yellowish-brown and become twisted inside their not-yet-capped cells. If an EFB outbreak is small enough, adult bees may be able to remove the diseased larvae and, with it, the *Streptococcus pluton* bacterium. In more advanced cases, some beekeepers step in with antibiotic treatments, but any honey stores inside a hive during treatment aren't safe for people to eat.

Adult bees have their own problems, with most of them resulting in bee diarrhea. Sometimes the poor bees barely make it outside to relieve themselves, and the front of their hive and the ground just around it are splattered with the yellowed, streaky evidence.

Stressed-Out Bees

Honeybees can get as run-down as the rest of us, and as with people, stress can weaken their ability to fend off illness. In many cases, the bees that, despite their natural enemies, do manage to keep going are expected to complete rather punishing pollination circuits. Their hives are loaded onto trucks and driven from state to state, so that the bees can pollinate apples, apricots, almonds, and more. In a typical season, bees might log miles from Florida to Texas, from Texas on to Arizona, from Arizona to California, and beyond. That's a lot of wear and tear on the honeybees, and it looks as if it's finally catching up with them and, by extension, us . . .

The Biggest Threat Yet

Within the last couple of years, honeybees have had to face yet another adversary—what researchers are calling Colony Collapse Disorder (CCD). So far, its effects have been devastating, accounting for the loss of between 45 and 90 percent of some beekeepers' hives. The syndrome causes seemingly healthy colonies to disappear nearly overnight. In CCD-affected hives, honey stores and capped brood usually have remained intact, but most all of the bees—save the queen and a few pitiful handfuls of the very young—have vanished.

Presently, entomologists have more questions than answers, and while they're still searching for clues, CCD has popped up in at least thirty-five states and across Europe, Asia, and South America. Some theories include pesticide poisoning, some new or existing parasite or virus, or as the USDA suggests, a set of immune-suppressing factors such as drought, migratory stress, and poor honeybee nutrition "due to apiary overcrowding, pollination of crops with low nutritional value, or pollen or nectar dearth." In other words? For far too long, our honeybees have been ridden hard and put away wet.

Scientists are scrambling to find CCD's cause—and a cure. For now there are at least some promising developments. Jerry Bromenshenk is one of many researchers at work on the problem. Because a hive's buzzing can change under certain conditions, Bromenshenk has inserted microphones less than an eighth of an inch in diameter into the heart of honeybee colonies to digitally record the sounds they make. The sonic fingerprints or "signatures" of each of these recordings are then analyzed, and some interesting patterns have emerged. For one, when exposed to sublethal chemicals, honeybee colonies have reacted quickly, altering their sonic signatures in the same ways from colony to colony. The researchers were amazed to discover that they could actually identify the kind of chemical any hive of honeybees was exposed to by studying its sonic signatures. The same holds for honeybee colonies plagued by Varroa mites, and not only could researchers differentiate colonies with Varroa mites from those without them, but they could also distinguish low-level infestations from medium and high levels.

By sampling the sounds of thousands of hives, scientists are slowly building a library of the different sounds—and the sound signatures—bees make under all sorts of conditions. The hope is that, one day, beekeepers may have a new tool in their arsenal, one that Bromenshenk likens to Star Trek's famous "tricorder." With such a device, a beekeeper could "scan" his hives for signs of mites, chemical exposure, or disease, and help his bees accordingly.

* * *

That honeybees are neither bald eagle majestic nor pygmy rabbit cute hasn't helped their cause. Too much rough handling—and a general disregard for the honeybee—ultimately, will translate into higher prices for nuts, fresh fruit, and veggies at the grocery store and fewer kinds of available produce in general. But it's not too late to help the honeybees. That's why, even though fending off those parasitic mites, assorted pathogens, and any lingering, negative perceptions about *Apis mellifera* can be daunting, we diehard beekeepers will never give up. After all, beekeeping has long been about preparing for the worst, while always, always hoping for the best.

Part Two

A Beekeeper's Life

CHAPTER SIX

A (Brief) History of Beekeeping

We may never know exactly how it was that humankind had its first taste of honey. As I see it, some hulking caveman type was probably out hunting for food when he stumbled over a fallen tree. Curiosity draws him to peer into a large cavity along its trunk, and as he thrusts his hairy mug inside, he gets a beard full of peeved bees (and a good story to tell Ogg and the boys). Or maybe most of the bees have already vacated, leaving some stragglers and just enough of their sweet, golden honey to pique his curiosity. Somehow, anyway, people came to value honeybees and the honey and wax that bees make well before they realized how to look after the nests of wild bees living in the hollows of trees or deep inside rock faces. It would be a long time, too, before they would "keep" bees in man-made hives.

Death and Taxes

It's widely believed that the Egyptians were the earliest beekeepers, and they valued honey so highly that honey-laden comb was regularly presented as an offering for the afterlife. The living also relied on it—as a sweet treat and to treat wounds. "There is a lot of documentation about their use of honey as a medicinal for cuts and burns, and it does have very good antimicrobial properties. It was widely used as an ointment," according to Gene Kritsky, a biology professor at

Beekeepers were depicted in Egypt's Tomb of Rekhmire. *Gene Kritsky, College of Mount St. Joseph*

the College of Mount St. Joseph. A few years ago, while visiting the Egyptian Tomb of Rekhmire, Kritsky studied the original reliefs on the tomb's walls. "There are several areas at Rekhmire that show different provinces paying taxes in honey as well," he says.[1]

But the tomb walls showed much more than that, including stacked beehives sitting atop a table. "The interesting thing about those particular beehives is that they really don't look like any of the other tomb drawings of beehives. They look almost like an oven," he adds. Actually, the ancient apiculturists hived their bees in long, cylindrical tubes, which were then horizontally stacked on top of one another. The beekeepers are depicted accessing each hive cylinder from the back and handling smoky incense burners around their bees.

Kritsky and other Egyptologists believe the beekeepers understood that using smoke could calm the bees. In a scene

1 If only the IRS would accept payment in honey! Last year's tax bill would've amounted to something in the neighborhood of 1,474 pounds of honey, but considering that I only keep a few hives of bees, I guess I'd better stick to writing checks . . .

from a Fifth Dynasty sun temple, which predated Rekhmire by about 900 years, a beekeeper is shown hoisting what looks like the smokers used in ancient Greece. "The part that shows the worker allegedly smoking the bees is broken, so you can't see what he's actually doing," Kritsky explains. "It's not entirely certain whether he is smoking the bees, but the actual hieroglyph symbols that are in that scene are consistent with beekeeping, meaning to 'quiet' or to 'loosen.'"

To harvest honey from the cylindrical hives, it's likely that the beekeeper would have opened the back side of one of the tubes, used some smoke to quiet the bees, carefully loosened the honeycomb, and then removed some of the disk-shaped sections. After that, they'd mash the honeycomb parts into large mounds and pour the honey off into containers to be sealed.[2] Unlike the golden honey I periodically collect, the Egyptians' honey ranged from amber to bright red, and lest you think such coloration was merely the result of artistic license, think again. Egyptian writings refer to red and amber honey, and after color correcting a photo of one of the reliefs, Kritsky says, "It shows the beekeeper and his hands in a worship position honoring the hive, and then the upper panel shows a bee worker pouring honey into containers for sealing. And when I photo corrected it . . . the honey came out

2 I never before realized how much I'd had in common with the ancient Egyptians. The first time I took any honey from my hives, I cut sections of honeycomb out—larval bees and all, I'm afraid—tossed everything into a large mixing bowl, and mashed it all up with a wooden spoon. Next, to separate the wax and God-knows-what-else from the honey, I strained the glop through some cheesecloth and proudly bottled the slightly off-colored, cloudy stuff. (The good news? I used most of that batch of honey myself, but I may have foisted at least one of the jars on my parents. Sorry, Mom and Dad!)

red, which is typical of honey from the desert, and the amber honey is more from the Delta region."

Sketch of a partial beekeeping scene from the Tomb of Ankh-hor.

There are often-told tales of archaeologists coming across 2,000-year-old pots of honey that were still intact and quite tasty, but the jury's still out on that.[3] "As far as that story of people opening up the pot and pulling up honey—and I think the story goes that there was even a fetus in there or a hand or something—that I can't verify. Maybe it was an Egyptian urban legend or one of the first urban legends ever," Kritsky suggests.

There is still a lot we don't know about those ancient Egyptian beekeepers, but there may be at least one or two other beekeeping scenes out there to help us fill in the blanks. Purportedly discovered during the 1930s, one scene from the Causeway of Unas is said to be similar to the one from the Fifth Dynasty sun temple, and Kritsky says, "There is possibly another beekeeping scene in Ankh-hor near the Tomb of Ankh-hor—the Tomb of Harwa—that's being excavated by the Italians now."

3 Since bacteria can't grow in it, properly cured honey—with its low moisture content and low pH—can last indefinitely.

Other Early Beekeepers

There is evidence that beekeepers in ancient Greece and Rome also relied on horizontal hives, but the Romans, at least, experimented with other hive shapes, materials, and beekeeping practices. They sometimes housed their bees in hollow logs or wicker "pots" made from woven branches, and they were even known to feed the bees from time to time.

Throughout Europe's forests, people hunted for wild nests of bees high in the treetops. As early as AD 1000, Russians cut small footholds in bee-filled trees in order to make climbing up them to harvest their honey much easier. By the 1700s, sundry picks and rope, slings, and even spiked footwear were also widely used in Russia, Poland, Germany, the Baltic region, and outlying lands.[4] After making their way up to wild bee nests, beekeepers would sometimes cut small access doors into the host tree—ostensibly to make subsequent visits less labor intensive—and to help bees overwinter in very cold climates, they also would tie a layer of straw around the hive portion of the tree's trunk. As for countries with more rock formations than trees, people tended to wild rock cavity nests in much the same way.

Actor Henry Fonda was a beekeeper. So is home-and-garden guru Martha Stewart, who has been keeping bees since the 1970s. Other famous beekeepers have included Brigham Young, Leo Tolstoy, Sir Edmund Hillary, and the Ukraine's Viktor Yushchenko.

4 I've long admired the Germans' ability to smash smaller words together to create one very big, more specific one. Forest beekeeping was known as *Waldbienenzucht*, which roughly translates to "forest bees breeding."

Slings and ladders were among some early beekeepers' most important tools.

In time, people decided to work with bees closer to home. In exotic locales such as Greece, Turkey, and Iran, hives were built into walls, and in heavily forested lands, treetop nest tending gave way to hanging, log beehives that could be raised and lowered as needed. Early beehives destined to remain on the ground were crafted with local materials. For example, those in Spain and Portugal were made out of cork, while those in countries such as England, France, and the Netherlands were more often fashioned from wood or clay-covered straw. Even with beekeepers' constant tinkering across every continent, it would be a long time coming before beehives looked anything like the ones in the modern beekeeper's apiary.

America's Honeybee Patriots

Lore has it that a twelve-year-old Quaker girl—along with several hives full of honeybees—"saved America" during the Revolutionary War. Tammy Horn, senior researcher apiculturist at Eastern Kentucky University and the author of *Bees in America: How the Honey Bee Shaped a Nation*, originally unearthed the story from a 1917 issue of *American Bee Journal*, but scholars haven't yet been able to verify whether or not the event actually took place. Even if it is just a tall tale, it's certainly a remarkable one.

It was the summer of 1780 when General George Washington and his "ragged, half-starved soldiers" were camped just outside Philadelphia. An attack on the Revolutionaries was imminent, and the original text asserts, "Man after man had risked his life trying to get their secret, but so far no one had been able to give Washington the important news without which he dared not risk his small force in battle."

Meanwhile, as she walked down a path lined with beehives, the local girl encountered a badly wounded Revolutionary soldier who did manage to uncover the British army's plan. He rode on horseback and entreated her to get this message to Washington: A large army was coming on Monday. Because the British soldiers weren't far behind, the girl mounted his horse, grabbed a large stick, and before galloping away, "with a smart blow, she beat each hive until the bees clouded the air." The bees subsequently attacked the Redcoats, and the narrator claims, "If they had been armed with swords the brave bees could not have kept the enemy more magnificently at bay."[5] Upon receiving the girl and hearing her story, Washington purportedly said, "Neither you nor your bees shall be forgotten when our country is at peace again. It was the cackling geese that saved Rome, but the bees have saved America."[6]

Americans consume about 1.31 pounds of honey per person annually.

5 That wouldn't have been the first time honeybees were used in warfare. Hives of bees have been dropped on and hurled at advancing forces just about everywhere, including Greece and Rome, across Europe, and in Africa, too. During the Civil War, the South shot through some hives to beat back the North, and while the Vietnam War raged, Vietnamese guerrillas fitted some beehives with trip wires that, when triggered, would disturb the bees along the oncoming Americans' path. (Sadly, the honeybees have never had a say in their enlistment.)

6 Interestingly, some beekeepers who came long after George Washington believed if they didn't move each of their hives at least an inch or two on February 22—Washington's birthday—all of their honeybees would die.

Before the movable frame hive came around, honeybees might be kept in a woven, straw skep like this one.

* * *

Apis mellifera had officially arrived in North America by the 1600s, and it was the English settlers who systematically imported honeybees. Typically, the bees were kept in straw skeps—those coiled containers most of us still conjure up when we hear the word "beehive." Bottomless and hollow on the inside, the skep had one bee-sized entry point toward its base, making it nearly impossible to check the hive for signs of trouble. "Straw skeps were quite vulnerable to things like wax moths. I write at one point about German beekeepers who had one hundred [skep] hives or so and then lost them all in two or three weeks because of wax moths," Horn says. Because bees also quickly became cramped inside the small skeps, they frequently swarmed—leading to the creation of a new, additional queen and the invariable loss of at least half of one's hive population. Another particularly nasty aspect of the use of skeps was that, in order to harvest honey from them, beekeepers often killed the bees contained inside.[7]

Lorenzo Langstroth was an American beekeeper who popularized the movable frame-style hive.

To make up for those frequent losses, searching out and collecting honey and bees from wild nests was deeply ingrained in colonial culture; just as they depended upon chickens and other livestock, the colonialists relied on honeybees, too, to help them get through hard times. In fact, bee trees were so prevalent in early America that honey hunting very nearly became its own sport. Frequently engaged in good-natured competition with one another, bee tree trackers competed to see who among them had the keenest eyesight, the swiftest feet, and the easiest time navigating the wilderness. But in light of the South's multiplying cotton plantations and the industrialization of the

7 Bee-filled skeps were routinely crushed, immersed in water, or "brimstoned"—poisoned with sulfur fumes—but there were at least some apiculturists who took the time to attempt a transfer of bees from one honey-filled skep to another, empty one.

North, forested land was harder to come by, and the pastime gradually died back.

At about the same time, beekeepers began to experiment with new "movable frame" hives, which allowed them to pull out and examine entire sections of honeycomb at a time. European inventors had started to tinker with these box-style hives as early as the late 1600s and 1700s, but an American beekeeper named Lorenzo Langstroth is one of the best known.[8] Today, most modern hives—mine included—are essentially Langstroth hives or slight variations on his original design.

From a distance, the modern beehive's wooden boxes look like dresser drawers missing their knobs. Really, though, each of these "drawers" has neither knobs *nor* bottoms. In beekeeping parlance they're known as "supers," and stacked one on top of the next, each individual super usually contains 10 rectangular frames full of honeycomb and bees.[9] Langstroth's original hive—and the hives of several inventors including Russia's Peter Prokopovich and England's Augustus Munn—took into account the idea that bees are very particular about their space.[10] If there is too much room between individual frames in a super—say, more than a half inch—honeybees will automatically close the distance by filling it with extra honeycomb. And if there is extra space between the outside frames and the inside wall of the super? Bees will fill that, too, with

8 Langstroth penned the classic *The Hive and the Honey Bee*, and he frequently made the rounds in the apiculture journals of his day.

9 A super, when filled with honey, can weigh 40 to 50 pounds, so to help lighten my load, I use supers built to hold only eight frames—but they're still really heavy!

10 In the early 1800s, Peter Prokopovich created a cabinet-style beehive that allowed for the right amount of space between its honeycomb sections, but not between the outlying sections and the walls of the hive, and during the 1830s and 1840s, Augustus Munn experimented with beehives which had just a half-inch space between and around its frames.

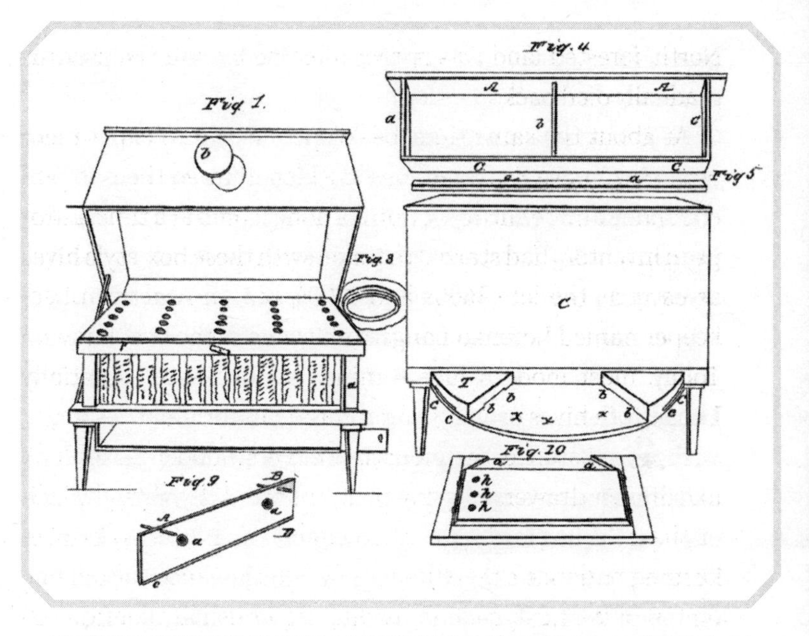

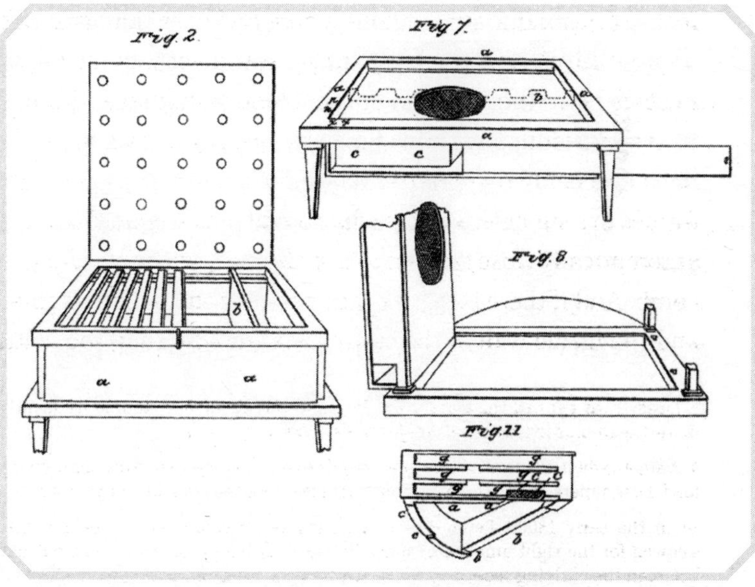

These patent sketches detail Langstroth's movable hive invention.

more honeycomb or, sometimes, with propolis, making the removal of single frames for inspection messy and difficult.

In Langstroth's hive, there was just barely enough room for bees to pass between the individual frames and between the frames and the top, bottom, and sides of the hive box that housed them. That meant they wouldn't feel the need to fill in any extra areas, and so beekeepers would be able to inspect a hive frame by frame without doing damage to the honeycomb or the bees. In his 1852 application for a U.S. patent, Langstroth enumerated several of the benefits of his creation, such as providing better protection from wax moths and shielding bees from extreme heat, cold, and dampness. It also enabled the beekeeper to easily collect honey and "perform all necessary operations without injuring a single bee." Beekeepers might also escape uninjured, since the hive allowed even "the most timid to remove the surplus honey without danger to themselves." It must be said that Langstroth was a gentle, humble man who never set out to steal others' thunder. When he wasn't fiddling with his honeybees, he served as the principal at a couple of girls' schools and sometimes as a church pastor, and he did his best to give credit where it was due. The guy even died with class.[11]

11 A. I. and E. R. Root's 1917 edition of *The ABC and XYZ of Bee Culture* notes:

Mr. Langstroth died Oct. 6, 1895, at the Wayne Avenue Presbyterian Church, Dayton, Ohio, where he was preaching that day. Before he began, the pastor of the church, Rev. Amos O. Raber, moved the pulpit to one side and placed a chair in front where Mr. Langstroth could sit while speaking, for his enfeebled condition would not permit him to stand. After a few requests for prayer on the part of the congregation, he said: "I am a firm believer in prayer. It is of the love of God that I wish to speak to you this morning—what it has been, what it is, what it means to us, and what we ought—"

His daughter, Anna L. Cowan, who was present, thus describes the last scene:

That said, Langstroth is also often credited with inventing an ingenious device known as the queen excluder, but Prokopovich, a Frenchman named Abbé Collin, and a few others independently came up with variations on the same idea much earlier than he did. Looking like prison bars in some Lilliputian lockup, a queen excluder is simply a thin metal grating that is placed one per hive between the brood area where the queen and her attendants live and the honey supers where the workers store their surplus nectar. The excluder's bars are spaced closely enough that the workers can pass through, but the queen cannot. A simple enough idea, but combined with the movable frame hive, it would revolutionize beekeeping; apiarists would finally be able to easily harvest pure honey without hurting honeybees or ruining their homes. That made apiculture more profitable and, I think, more fun, which would certainly help in the years to come.

Putting the "Bee" in B-52

The need for beekeepers dramatically increased during times of war—especially during World War II. "It was a way to vary a very bland diet during World War II rationing, and of course, it was a patriotic thing to do, too, because there was a constant demand for beeswax," Horn says. From waterproofing

---

"As he finished the last word he hesitated; his form straightened out convulsively; his head fell backward, and in about three minutes he was absent from the body, at home with the Lord.

"There was no scene of confusion in the church. Tears were running down every cheek, but there were no screams, no loud sobbing. As one person remarked, 'Heaven never seemed so near before. It seemed but a step.'"

As the sign at this New York City grocery suggests, sugar rationing was just another fact of life during World War II. *Getty Images*

tents and metal casings for bullets to lubricating mechanical parts, beeswax was instrumental in the war effort. It was even smoothed onto planes to help them cut through the air more efficiently. The commodity was so desirable that some beekeepers were permitted to defer the draft in order to tend their stateside hives, but by and large, more than a few of America's beekeepers were manning the front lines.

For those beekeepers remaining at home, there were other perks. Thanks to a federally sponsored initiative, beekeepers had more access to lumber, and because it was often necessary to feed the bees in order to get them through the winter and to help build up spring bee populations, beekeepers were entitled to extra sugar rations—at least in theory. "Many administrators were not educated about the special requirements of beekeepers, and they ended up exercising their power a little too judiciously, with unfortunate consequences for bees and their keepers," Horn explains. As a result, beekeepers weren't able to make up syrup to feed their bees during rough winters, and they lost their hives. That would be the start of a long-term decline in the number of hives and beekeepers in the United States. It would be a decline that would continue all the way to the present, but with any luck, that might be about to change, at long last.

CHAPTER SEVEN

Got Bees?

You might be surprised to know that honeybees can be shipped through the U.S. mail. To stock my own hives and become a bona fide beekeeper, I would need literally thousands of honeybees, and strange as it sounds, beekeepers buy their bees by the pound. From talking to other beekeepers and perusing the tiny line ads in the backs of my *Bee Culture* magazines, I knew I could order several kinds of honeybees from states down South. I picked a little apiary in Georgia that appeared to specialize in Italian bees and dialed their number. It was late February—soon enough, I hoped, to secure my very own shipment. Although I hadn't exactly expected a perky receptionist to answer, I wasn't entirely prepared for the elderly lady who did.

I could hear a TV blaring in the background, so I did my best to speak up, but it wasn't enough. She laid the receiver down with a clunk, shuffled away to quiet an old episode of

Family Feud, and then made her way back to the phone. Her son was the one I'd need to speak with about the bees, so again, I heard the handset plop on its side, followed by more of her scuffling and then the slam of a screen door. Minutes passed, and I thought she'd forgotten me. If I hadn't wanted those Italians so badly, I might've given up. Instead I pressed my ear closer to hear Richard Dawson shout, "Survey says!" followed by a game show *Ding!* and the requisite applause. The screen door banged once more, and here she was: "Ah think Earl's list is full up, but you tell him Mama said to put you ohn it. Mama said!" Once I had Earl on the phone, that's just what I did. He'd told me he was all out of bee packages, but he must really love—or fear—his mama. My bees arrived in the early spring.

* * *

Prices do continue to rise, but three pounds of bees and a queen can cost roughly $40 or $50 plus shipping.[1] Generally, when a cage full of live bees shows up, the post office notifies its recipient with uncharacteristic alacrity.[2] The bees usually come in a wooden box, and so that plenty of fresh air can circulate through, its sides are screened with wire mesh. This arrangement allows for an up-close view of thousands of honeybees.

1 I've never tried to count them myself, but most people say there are, give or take, about 4,000 bees to a pound.

2 A rather rattled postmaster once phoned a beekeeper I know at 2 a.m. to let him know his bees had arrived—and to ask if he would please come and get them at once. Instead, the beekeeper suggested they tuck the cage into a quiet corner somewhere until he could come for them—at a more reasonable hour.

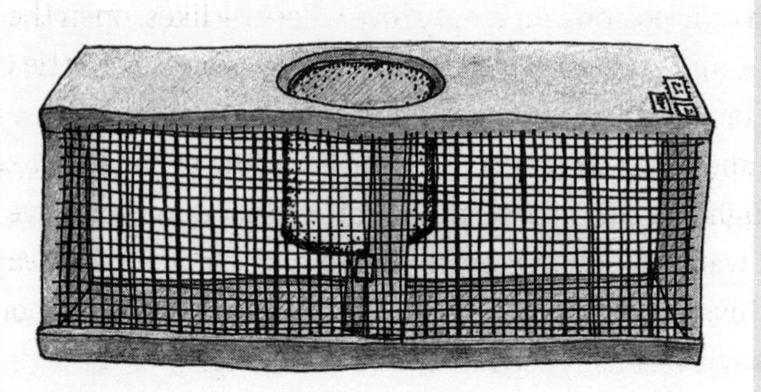

Aside from a can of sugar syrup and a small queen cage, the typical honeybee shipping container also holds a few pounds of honeybees.

Now, holding a cage full of bees is quite something. Not only are they much heavier than one might expect, they also give off an unusual amount of heat while crammed together. In the center of the cage's cluster of bees is a large can of sugar syrup to tide them over until they're free to collect nectar for themselves. Fastened to the top of the cage near the syrup, a queen bee and some attendants are further isolated in a tiny cage of their own. The insect equivalent of cranky kids enduring a long car trip, a bee package in transit can be quite loud. To soothe bees on the go, a beekeeper might use a water-filled spray bottle to mist through a cage's screened sides. Each spray of the bottle will soften the bees' sounds, at least for a little while. One unfortunate albeit inevitable consequence of purchasing out-of-state bee packages: Scores of the insects will die before they can be properly hived.

Mercifully, freeing one's bees is easy enough. The beekeeper need only position an empty hive where he likes, open the bee crate, and dump them inside. As for the queen, her little cage can be placed on top of the frames in one of the supers, and she and her new charges will get to know one another soon enough. The last step? Replacing the inner cover and hive top and watching the stragglers flying about make their way to the hive entrance. In just a couple of days, one's new honeybees will be busily setting up housekeeping.

Building a Better Bee

These days, when beekeepers place their orders for honeybees, they don't have to stick with the same old Italians. Thanks to a decade's worth of research by entomologists, beekeepers now have access to bees bred to have some very handy qualities, which make keeping a hive healthy a little easier. Carrying a genetic trait that most bees, to some extent, likely already have, some bees actually protect themselves from the parasitic Varroa mite quite well. Building on the work of Marla Spivek, USDA Agricultural Research Service entomologists Jeffrey Harris and John Harbo uncovered a genetic trait known as "Varroa Sensitive Hygiene," or VSH, and the bees lucky enough to express the gene will fare much better than those who don't.

Here's how it works: VSH bees are real neatniks who notice when mite families have formed inside honeybee brood cells. Typically, adult female mites will invade a brood cell just before the nurse bees are able to cap it. So they won't be detected, they hide beneath the larvae, and once the nurse bees *do* cap the cells, the female mite lays several eggs, which will hatch and, ultimately, feed on the live, larval bee. Their feeding doesn't usually kill the bee, but it can cause the bee to be weakened, diseased, or even crippled. When it's time to emerge from the brood cell, the mother mite and her female offspring will hitch a ride on the back of the bee, and the cycle continues.

To fix this very serious problem, bees with the VSH trait actually tear off each infested cell's beeswax cap and scoop out the cell's contents. "This interrupts the mite's family and, basically, kills her offspring and interrupts her reproductive cycle. If you do that all the time, the mite population eventually falls," Harris explains. Purging any mite-infested cells before the fully formed mites are able to emerge may badly disrupt the Varroa mite's life cycle, but it isn't a terrible hardship for the honeybees. They simply clean up the previously infested cells, and the queen comes around to lay new eggs inside them.

Interestingly, the bees remove bee larvae only in cells containing whole families of mites, and Harris and his colleagues still aren't sure why that happens. "One possibility is there could be something different about the host pupa that's being fed on. There might be distress signals or disease conditions that the mite family brings to them—like viruses—that a large mite family causes to be different than a single mite that doesn't lay eggs that bees can smell," he suggests.

Whatever the reason behind the bees' hygienic behavior, in time the mites could wise up and find a way to circumvent it.

"If you think about nature, it's always an arms race between predator and prey and pest, and they're always evolving. You know, the antelope get a little faster, and so the leopards then have different tactics. Things change constantly," Harris says. "So one thought we've had is, let's get a second resistance trait—at least one other and maybe two others that we can find and incorporate into a mixed stock." They're hot on the trail of one of those extra traits now. Known as "nestmate grooming," adult bees exhibiting the behavior remove Varroa mites from the backs of other bees. Harris says, "They'll grab Varroa mites off of the bodies of their sisters and bite them and pinch them and drop them out of the nest. . . . So, if you have two traits in one bee, then it's much harder for the mite to get resistance to both mechanisms."

But the new-and-improved bees don't come cheap. Two pounds of bees and a hygienic queen usually cost about $70—almost twice the usual price. But if beekeepers relying on specifically bred bees rarely (or even *never*) lose their hives to mites, forking over that extra money seems well worth the cost. Down the road, beekeepers may even have access to honeybees with multiple, positive genetic traits. "Mainly there's three or four traits we go for," Harris continues. "They have to be resistant to Varroa, resistant to tracheal mites, have low defensive behavior—meaning they don't sting very much—and then high honey production. And you select for all of those at the same time."

Besides keeping bees expressing the VSH trait, there is one other comparatively low-tech weapon against the dreaded Varroa mite. Placing a wire-screened board at the very bottom of a beehive allows air to flow in and any mites the bees may dislodge to fall through. And finding their way back to the hive entrance or through the screened bottom board is nearly impossible for the tiny mites. (So, ha! Take that, you yucky mites.)

A Beekeeper's Honey-Do List

To properly look after beehives during peak nectar flows, the beekeeper necessarily becomes almost as busy as his fuzzy flock. But long before he lifts off a single hive top, a beginner will likely fortify himself with coveralls, elbow-length gloves, and apiculture's signature hat and mesh veil.[3] The single pieces, when worn together, can make just about anyone sweat in rivulets. Add in the 90-degree summertime temperatures, and one fairly roasts. It's not that the manufacturers of all this gear haven't tried to make their customers more comfortable. Although beekeeping coveralls are usually made from heavy cotton cloth or, sometimes, a cotton-and-polyester blend, they are, at least, a light-reflecting white. Of course, with so many zippers, buttons, and Velcro tabs, the full-body suits allow neither bees nor fresh air to penetrate. A beekeeper's gloves, with their heavy canvas or leather "hands" connected to cotton or, sometimes, nylon sleeves, add an extra layer of heatstroke-inducing protection.[4] As to the veil, there are all kinds. Some

3 More experienced beekeepers hardly ever bother with coveralls or even gloves, but Nervous Nelly that I am, I feel more comfortable with them on. I even duct-tape any bee-sized entryways in my sleeves and pant cuffs to discourage curious bees from crawling inside and getting stuck.

4 Some beekeepers swear by nylon, because it's thought to be much too slippery for curious bees to climb.

of the lighter ones look like fencing masks attached to soft, stiff hats. Others are meant to fit over the top of hard, plastic pith helmets. No matter the style, none of the fine mesh nets is too unbearable.

Once fully protected, the beekeeper, lit smoker in hand, is ready to see if his bees have enough storage space for all of the nectar they've been carrying in. So as not to disturb the bees too much, it's best to approach the hive from behind. A few puffs from the smoker at the hive entrance come next, and then one lifts off the hive top to reveal the inner cover, a thin board with an oval-shaped opening in its center. It's the only thing separating the hive's interior from the rest of the world. A hint of smoke through the hole pushes the bees a little farther down into the hive, so now's a good time to remove the inner cover.

But that isn't always as easy as it sounds. Because bees often seal any cracks they find with propolis, separating an inner cover from its super can require real leverage. That's where the hive tool comes in. It looks like a miniature crowbar, and every beekeeper carries one to pry open his hives, separate stuck supers, and free up individual frames from one another. Unstick a hive on a scorcher of a day and the propolis stretches out like taffy to make an uncommonly sticky mess.

As long as the bees stay relatively calm, one can look through frame after frame of honeycomb to see whether the queen is healthy and laying, if the bees have been able to store lots of nectar and pollen, and whether they're running out of honeycomb cells in which to store these goodies and raise their brood. Bees that are pressed for space often will choose

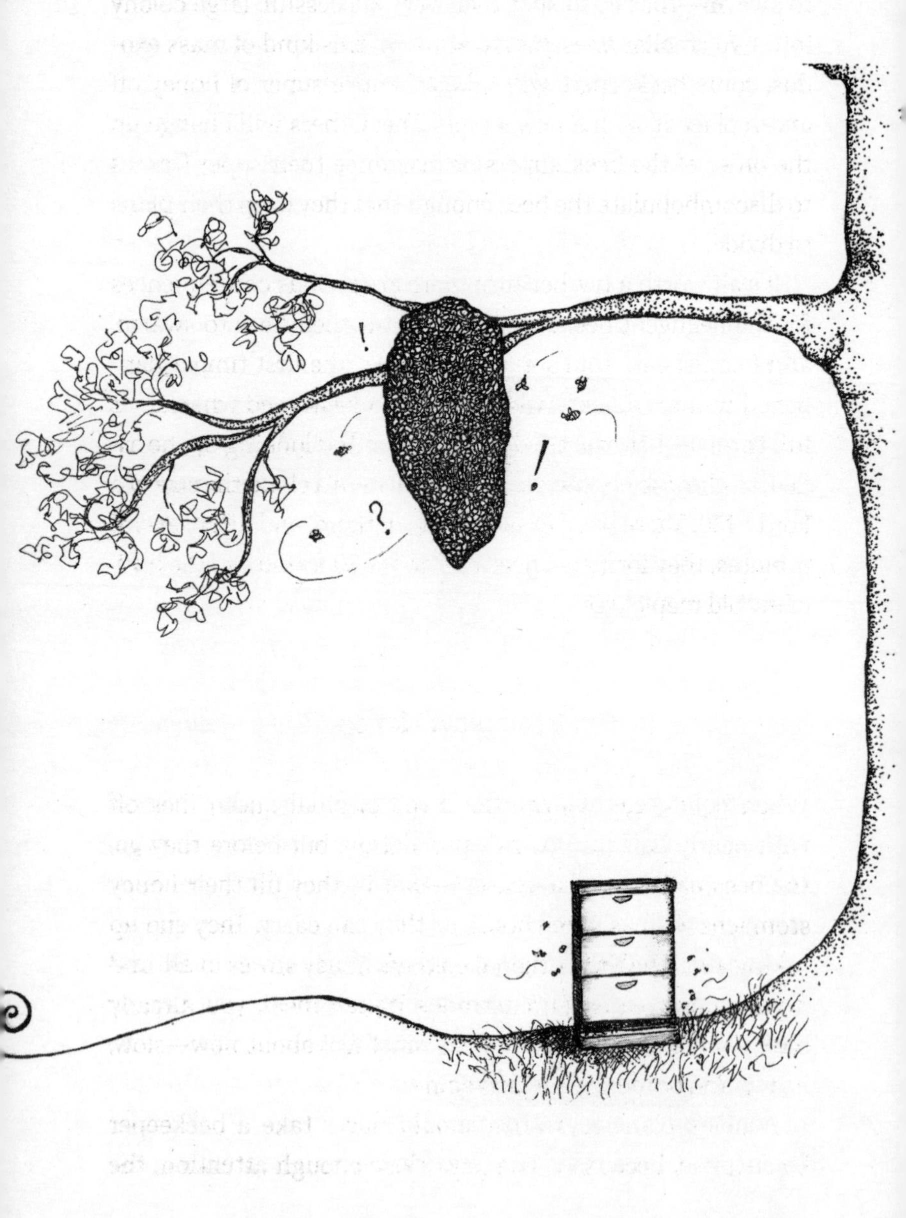

to swarm—that is, to split their very successful, large colony into two smaller ones. Just to stave off this kind of mass exodus, some beekeepers will take an entire super of honey off and replace it with a new, empty one. Others will change up the order of the bees' supers or rearrange their super frames to discombobulate the bees enough that they drop their plans to divide.

It's all worth a try, because there are serious consequences for the negligent beekeepers who allow their bees to swarm, and I can speak from sorry experience. The first time it happened to me, I'd heard what sounded like a weed whacker at full throttle just over my head. I remember looking up, horrified, to glimpse my wayward honeybees in a cloud the size of a Ford F150. There were easily 30,000 of them, and after several minutes, they formed a neat ball about 20 feet up on the limb of my old maple tree.

Bees in the Trees

When honeybees swarm, the hive's original queen flies off with nearly half the colony's population, but before they go, the bees pack their "suitcases"—that is, they fill their honey stomachs with as much honey as they can carry. They end up taking about half the original colony's honey stores in all, and if you've ever eaten an enormous Italian meal, you already have some idea of how the bees must feel about now—slow, heavy, and bordering on food coma.

A hive's plans to swarm should never take a beekeeper by surprise, because if one pays close enough attention, the hallmarks of an ensuing swarm are apparent. Weeks before its big departure, the colony must rear a new queen to take the old one's place in the hive, so a beekeeper finding lots of queen cells—usually stuck along the bottoms of honeycomb frames—can be fairly certain that something fishy's going on. Timing is everything for the new, would-be queens, and when one or more emerge, they'll chew through any queen cells that haven't yet hatched, thereby eliminating the competition before it can emerge. And if more than one emerged queen is out running around? The queens battle it out, and the last one standing will stay behind to head up what's left of the original colony.

In the meantime, the old queen, and the thousands of swarming bees accompanying her, cluster tightly together in a temporary location while scout bees head out to find new digs. Fairly recently, biologists at Cornell University discovered that this process is actually quite democratic, and just as honeybees tell one another about the availability of nectar via their waggle dances, location scouts, likewise, will dance to let the other bees know about potential new home sites. The dances that are the liveliest and most enthusiastic garner the most attention, and soon, other scouts head out to see if a particular spot lives up to its hype. If they're impressed enough, *they'll* join the first scouts and dance on behalf of the potential nest site. Even though there may be several groups of scouts recommending many different locations, the group that attracts the most bees the soonest usually wins. Most intriguing, fifteen seems to be the magic number. Once fifteen or more bees agitate for a move to the same new hive location, word goes out that the decision has been made. In no time, the bees warm up their wings and fly off.

* * *

A swarm of honeybees in this special state of limbo can look threatening, but they're often quite docile. Fully aware of this, some beekeepers try to impress onlookers by approaching a swarm with the sparest of tools. A box or a bucket and a broom suffice for bees stuck in someone's eaves. An extension ladder might come into play for bees not too far up a tree. Loppers and a large tub work well for bees settled in low-growing shrubbery; the beekeeper need only snip off the branch, lower it into his container, and slip away with bees and all.

Collecting a swarm isn't always so cut-and-dried, though. Sometimes one's bees will have selected such a high-in-the-sky hangout that they can't be rehived safely. There are "remedies" for these situations, but most beekeepers agree that they're more quaint than they are effective. Case in point, rather than feel utterly helpless, there are beekeepers who, wielding hand mirrors, have tried to direct a swarm's movements by bouncing sunlight near the cluster of bees. The thinking there is that the bees will follow this faux sun anywhere the beekeeper wishes. And then there are others who've gone so far as to turn the hose on a swarm just out of reach, hoping the drenched, dislodged bees will relocate somewhere a little closer to earth. But it is never, ever so easy.

Better than having to catch a swarm is making sure one's bees don't swarm in the first place. Good beekeepers have long tried to do just that, and in *The Sacred Bee in Ancient Times and Folklore*, author Hilda M. Ransome attributes the following to the *Syriac Book of Medicines*: "'To prevent bees leaving—Heat water and wine and rub the mixture on the hives, or fumigate them with burnt asses' dung.'" That I've never tried, but in addition to making sure they have ample room in the hive, I regularly look for queen or "swarm" cells, and if I find them, although I hate to do it, I crush them with the edge of my hive tool. Really, though, once the hive decides it's time to swarm, there's not much one can do to change its collective "mind."

If You See a Swarm

Swarming bees sometimes pick the weirdest place to land. They've been spotted on the outsides of mailboxes, on the legs of shopping carts, and on storefront signs, to name a few. I think it's good luck to find a swarm, and should you be lucky enough to see one clinging to the nearest telephone pole or maybe the awning of a building, you needn't do much of anything. During the location scouting process, a swarm usually only lingers a few hours or, sometimes, a couple of days before moving on. But if the bees are in a really inconvenient spot—or if you or someone nearby is allergic to them—call the nearest beekeeper or your animal control department for help.

To Bee or Not to Bee

Perhaps, by now, you've begun to imagine yourself, bee smoker in hand, tending to your very own beehives. You would surely become the Marlin Perkins of the beekeeping world, charming friends and acquaintances with droll accounts from your apiary. Your mailbox, consequently, would overflow with dinner party invitations. Your bees would never stray, and you'd need a backhoe to move all those jars filled with the purest, most toothsome honey ever collected . . . Well, fantasizing can be fun, but beekeeping is a big commitment. How can you be sure you'll have what it takes? And if you *do* don the beekeeper's veil, will you find sweet success or just end up getting stung? Here's a little quiz to help you find out.

1. My idea of awesome responsibility is:

   a. A pet rock.
   b. Training the family dog.
   c. Raising triplets.

2. While driving in your car, you happen to see a woodpecker dart by. You:

   a. Keep driving. You're probably running late.
   b. Slow down a little, craning your neck to glimpse him.
   c. Pull over to the side of the road and grab the binoculars you keep in the glove box for just such an occasion.

3. The neighbor kid accidentally breaks his ant farm, and now his ants are crawling all over the place. You:

   a. Shriek and stand on the nearest chair.

b. Head for the dustpan and broom.
c. Crouch down to see what they do. They're just so interesting!

4. It's 103 degrees in the shade, but your friends begged you to come camping with them. The mosquitoes are biting, but the fish aren't. How do you feel?

a. Duh! Totally miserable. How would *you* feel?!
b. Rather uncomfortable, but I'll be back in my own bed by this time tomorrow.
c. Happy as a clam at high tide. The view out here is spectacular!

5. My idea of a good workout is:

a. Workout? I need both hands just to lift the TV remote.
b. A brisk walk around the block.
c. Bench-pressing 250 pounds.

6. At the office, people say I'm:

a. All about the "big picture."
b. Pretty good at seeing the forest *and* the trees.
c. Obsessed with minutiae.

7. The surprise anniversary party you planned for your parents is turning out to be a disaster. You:

a. Throw a fit—then throw back a few gin and tonics.
b. Sneak into the kitchen to have a good cry.
c. Shrug your shoulders, and remind yourself that you did your best.

8. Your idea of a relaxing time is:

a. Spending a month in the Hamptons.
b. Swinging in a hammock while reading a good book.
c. Putting a new coat of paint on the garden gate.

9. Unfairly, you believe, you were slapped with a parking ticket. You immediately:

a. Gnash your teeth while paying the fine.
b. Make a few calls to see if you really have to pay it.
c. Stroll into City Hall, state your case, and get the ticket thrown out.

10. The last time you were stung by a bee, you:

a. Nearly died.
b. Can't remember what happened.
c. Swelled up a little, but it wasn't anything an ice cube couldn't fix.

SCORING

For every "A" answer, give yourself 1 point. For every "B" answer, award 2 points. "C" answers are worth 3 points.

(10–16 Points) Not to Bee—You might have become increasingly interested in these creatures, but you'll be better off admiring them from afar. Besides, there are plenty of other ways to give honeybees a leg up—and without all the stings and smoke.

(17–23 Points) Maybee—You're not quite ready to suit up, but you've definitely got potential. Doing a little extra reading and shadowing a beekeeper for a day or two will help you decide for sure.

(24–30 Points) Should Bee—A born beekeeper, you were made for this stuff. So what if you're all sticky and your back aches. It's all part of the deal, and you've never shied away from hard work anyway. These six-legged ladies will never fail to fascinate you, and—bonus!—you'll have months and months' worth of fresh honey while you're at it.

* * *

So just what do a broken ant farm, unpaid parking ticket, and benching 250 pounds have to do with beekeeping? I think quite a lot. No, taking care of bees isn't quite as big a responsibility as raising triplets, but it does require a fair amount of time, patience, and dedication. And about that woodpecker and those errant ants? It really helps if you're naturally curious about the world around you, and since facing thousands of bees for the first time can be a little disconcerting, it's also good to know how well you can keep your composure.

Beekeepers should be able to tolerate a little physical

discomfort, too. Being trapped inside coveralls, gloves, and a veil in the August heat can be unbearable, and at least once, a bee will make her way inside your suit.[5] Also, while you don't really have to bench-press like Sylvester Stallone, you *will* need to do some awkward heavy lifting—it's not easy to maneuver a fifty-pound super full of irritable honeybees. (Oh, and as obvious as this may seem, it's really important to know what kind of reaction you have to bee venom, because no matter how careful you are, occasionally you *will* be stung.)

Because trouble in a hive often starts out small, good beekeepers are also very detail-oriented. For instance, seeing more ants than usual hanging around the front of a hive can signify a weakened colony with fewer guard bees on duty to handle the intrusion. What's more, sometimes no matter how much attention you pay and how hard you try, you'll still lose a hive to Varroa mites, harsh winters, a failing queen . . . you get the idea. Think you can handle beekeeping's many ups and downs with aplomb? Since there's always something to do—repainting hive supers, putting new beeswax foundation in frames, or making sure the bees have all the space they need—you should also expect to stay busy and fairly close to home.

Finally, on that unpaid parking ticket: Sometimes beekeepers *do* have to fight City Hall. Many cities have already outlawed beekeeping altogether, and in some parts where beekeeping *is* permitted, it can take just a complaint or two

5 As if the sweat dripping into my eyes weren't bad enough, a single worker bee once gained entry under my veil and proceeded to walk back and forth across my forehead.

to convince city leaders to prohibit the pastime. Would you be willing to speak out on behalf of the bees at the next city council meeting? Unfortunately, you might have to—but plying them with honey really *does* help.

CHAPTER EIGHT

Pleasing the Bees

Whether you decide to become a beekeeper or not, there's a lot you can do to help any honeybees buzzing through your backyard. If you're a gardener with an abundance of several kinds of flowers and herbs, then you're already helping them some by providing much-needed nectar and pollen. One of the best ways to tell whether you're offering the right "menu" for honeybees and other pollinating critters is to take a walk outside and start looking for them. You'll probably notice bees choosing the dandelions and other weeds in your yard over some of your fancier hybrid flowers, but you might not understand why. Matthew Shepherd, a senior conservation associate with the Xerces Society for Invertebrate Conservation, has the answer: "On a number of occasions when flower breeders have been creating new varieties that are based on having double flowers or triple flowers or new colors, an unintended by-product is that the flowers produce less nectar and less pollen. They might look gorgeous for us, but they're not necessarily providing the food that the insects need."

So, if you do plan to plant a few extra flowers, you should choose native and old-fashioned "heirloom" varieties; since they haven't been overdeveloped, they'll provide a more reliable source of food. It's also a good idea to select flower colors that bees can easily see. Most bee-attractive flowers bloom in shades of yellow, blue, and purple—not red. Remember, bees don't see the same parts of the color spectrum that we do, and Shepherd says, "To a bee, red looks essentially the same color as green. That makes it very difficult for them to pick out a flower amongst the green foliage."

That may be, but why then are honeybees so often spotted working on blood-red poppies? Some red flowers—like poppies and blanketflower—have a very high ultraviolet reflectance, and bees have no trouble detecting ultraviolet. (By the way, while we can't *see* ultraviolet rays, our skin certainly reacts to them; too much UV spells sunburn.) If we *could* see ultraviolet light, Shepherd adds, "We would just see the poppies shimmering in the sunlight." We'd also be able to see all sorts of extra streaks and lines that, from a bee's view, surely look like the landing strips at an airport. They serve as nectar guides, pointing the way to sweet reward. The bees' ability to see UV has other advantages. Since they rely on the location of the sun to help them navigate, bees traveling on cloudy days might run into trouble. Thanks to those cloud-penetrating ultraviolet rays, though, bees have some notion of the sun's position in the sky, and they can still find the way to their favorite forage and back home again.

Because poppies are highly UV reflective, they really stand out to the foraging honeybee.

* * *

Should you like to become a real friend to honeybees, you'd better not be fussy about the white clover and dandelions growing in your lawn. In fact, by raising the cutting blade on your lawn mower a smidge, you can encourage these "weeds" to stick around and flower, so that area honeybees have access to a little extra free food. And if you live in a fairly out-of-the-way place with lax regulations about such things, relinquishing all or part of the lawn to taller weeds like milkweed, ironweed, and goldenrod is an even bigger help to bees, since these offer scores of tight flower clusters during late summer and early fall when the nectar flow is winding down.[1] But if you'd rather not let the yard go too wild, bees also love borage, sage, mint, thyme, lavender, and most other herbs as well as butterfly bush, daisies, honeysuckle, sunflowers, blackberries, raspberries, and the list goes on. They're also drawn to many types of trees—especially fruit trees like pears, peaches, apricots, and apples—but maples, willows, poplars, and locusts will do.

Other Guests You'll Meet

While inviting honeybees over to dine, you might see mason bees, mining bees, leafcutter bees, sweat bees—all sorts of wild pollinators, really—bellying up to the buffet. Between the solitary bees that nest in the ground, those that nest in old trees, and cavity-nesting bumblebees, there are over 4,000 kinds flying around. Unfortunately, just as honeybees have had their difficulties, so have our wild bees—only in their case, we don't have nearly as much information about them. "There are some bumblebees for which we have an increasing amount of very specific information about their decline, but for a lot of them, there's only anecdotal information. There's a lack of people looking and a lack of baseline data," Shepherd says.

1 In the fall my hives take on a most peculiar and pungent odor. It's partly sweet but mostly sour. At first I thought there was something terribly wrong, but in time, I realized I could attribute the stink to the bees' collection of nectar and pollen from local goldenrod. Asters, I'm told, can also cause a hive to smell a bit "off."

What researchers *do* know is feral bees don't have as much undeveloped habitat as they used to, and that, coupled with the spread of disease, has landed many of them on the "Red List" of North America's vulnerable, critically imperiled, or possibly extinct pollinator insects. A few kinds of bumblebees top the list, and, Shepherd notes, by making its way out of commercial greenhouses and into the wild, the Common Eastern Bumble Bee has had something to do with that: "*Bombus impatiens* is the main one that's being reared, and although as a species that bumblebee is not susceptible to some diseases and pathogens, it's able to carry them, and then these things have reached the wild-living species which are not as resistant."

If you want to go above and bee-yond (sorry) for any feral bees you may have attracted, you can provide a few types of nesting sites. About 70 percent of wild pollinators are solitary bees, which nest in holes they dig in the ground, so to better accommodate them, you need only leave some bare soil patches around for easier tunneling. The other 30 percent tend to nest in dead trees, but in a pinch, they'll settle in wooden nest blocks. To make one, the Xerces Society recommends choosing a block that's at least eight inches tall and drilling a series of holes between 3⁄32 and 3⁄8 inches in diameter. Any holes with a diameter smaller than ¼ inch should be drilled 3 to 4 inches deep. Those holes with diameters of ¼ inch or more should be drilled 5 or 6 inches deep.

Not that handy? You can bundle bamboo or other thick, hollow stems together instead. Just tie them with twine or tape them up with duct tape, making sure all the ends on one side are closed off in some way. Shepherd made one of

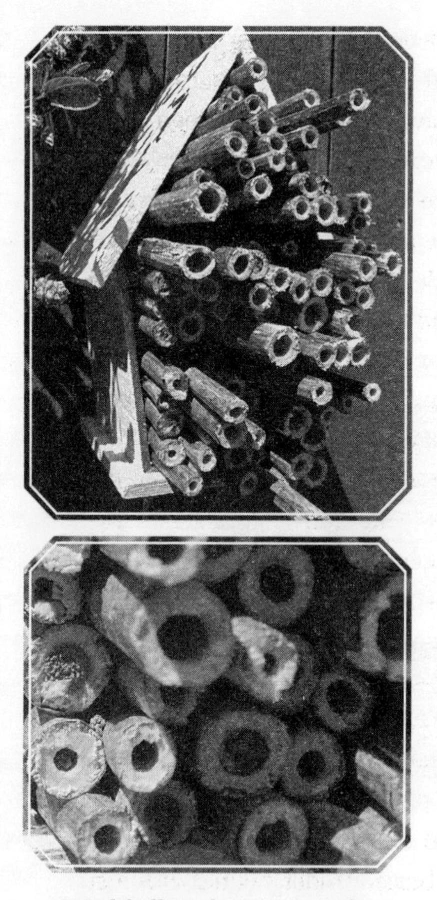

Dried, hollow plant stems make excellent "apartments" for solitary bees and wasps. *Matthew Shepherd*

his own solitary bee nests by grouping hollow teasel stems into a birdhouse without its front: "There's this little roof on it, and the stems stick out a long way." He mounted it about four feet off the ground, so that he could watch the bees coming and going from the comfort of a lawn chair. "You can watch the bees bringing the nectar and the pollen back. They tend to put the pollen in first, which is normally on their back legs or on the side of their abdomen, and so, they go backwards into the hole to give the pollen. And then, they come out, turn around, and go in headfirst to deposit the nectar that's carried in their crop," he says. Sometimes solitary wasps will move into a bee block, and they can be just as entertaining. "You can sit there and watch wasps bring in their caterpillar prey or their spider prey. I've seen a wasp fly past my nose carrying a katydid into the hole. You've got this whole nature documentary going on in front of your own eyes."

Providing shelter for bumblebees isn't much more complicated. Because they're social like honeybees, bumblebees make their homes in larger cavities—like abandoned mice or vole nests—and live together under one "roof." During the spring and summer, they make a colony and serve a single queen. "They all work to forage and bring food back to feed the offspring. They'll make little brood cells out of wax and little honey pots out of wax, and there might be fifty or sixty bees living in the colony at any one time, depending on the species," Shepherd says. But while honeybees make enough honey to sustain much of the hive throughout the winter, bumblebees do not. When winter comes, most bumblebees die, but the bumblebee queen goes into hibernation to emerge in the spring and begin the next colony.

Hosting a bunch of bumblebees during the spring and summer months doesn't take too much space. Wooden boxes that are 7 or 8 inches wide, deep, and high fitted with a ¾-inch, plastic tube "entrance" usually do the trick, but you don't necessarily need heavy-duty power tools to give bumblebees a place to land. Bumblebee homes can easily be made from a plastic foam six-pack cooler and a bit of tubing. They can even be as simple as a large teapot buried underground with just the tip of its spout exposed.[2]

A large, terra-cotta pot turned upside down also works. In time, bumblebees find their way through what used to be the pot's drainage hole. But part of the fun is spying on

2 But to keep the nest fairly dry, you might want to drill a few holes in the bottom of the teapot, and before you bury it, line the bottom of the hole with gravel to facilitate good drainage. If I were a bumblebee, I might even appreciate a little roof over the top of the teapot's spout to keep my home even drier.

the fuzzy beings as they go about their day, so I recommend using bumblebee bungalows fitted with observation windows. They come standard on most commercially available nest boxes, and the best way to position one of these is to bury most of it in the ground, leaving the roof accessible. Since most bumblebees are reasonably gentle, you can lift the top, wooden lid and peep down through the glass or Plexiglas window to see the colony firsthand.[3] Do that, and you'll probably find yourself spending a good amount of time doting on your guests. Now there's just one more detail that will make their stay more enjoyable—a bit of soft, upholstery batting, dried moss, or even dryer lint, so that they can "feather" their nest. "The thing to avoid is long fibers. So, don't put in a ball of wool, because bees have little tiny claws on their feet that can snag and get caught in long fibers," Shepherd says.

Home Sweet Home

Keith Bellinger, a fellow bumblebee appreciator, designed this cozy bumblebee house to sit aboveground. If you like a good weekend project, you might give it a try . . .

3 That doesn't mean *all* bumblebees are nice. There are a few species—including the very common American bumblebee (*Bombus pensylvanicus*), found in the Eastern United States and most states to the West, and the Sonoran bumblebee (*Bombus pensylvanicus sonorous*), living in Mexico and the Southwest United States—that can get kind of ornery, and since bumblebees' stingers aren't barbed like those of honeybees, they can sting multiple times without any dire consequences—well, at least not dire to them.

Well-placed ventilation holes, some cotton batting, and a PVC entryway make this homemade nest box perfect for bumblebees. *Keith Bellinger*

MATERIALS

- 5 feet of 1 × 6 untreated pine or hardwood
- Quarter sheet (2 × 4) of ¾-inch plywood
- 5 feet of 1 × 1 or similar sized molding
- 7-inch section of ¾-inch PVC pipe
- Screws or nails

TOOLS

- Drill with ⅞-inch and ¼-inch bits
- Saw
- Screwdriver or hammer

ASSEMBLY

1. Cut 1 × 6 board into lengths: front and back, 15 inches; sides and divider, 8½ inches.
2. Rip divider panel to 4½ inches by 8½ inches.
3. Cut plywood: bottom—12 by 15 inches, top—14 by 17 inches.
4. Fasten front and back panels to side panels, making a box 10 by 15 inches.
5. Fasten bottom plywood to box, leaving an extra 2 inches extending past the front panel to make a "porch" or landing pad.
6. Drill ⅞-inch hole in divider panel, 1 inch from one end and 1 inch from bottom.
7. Insert the divider panel 5 inches from one end, and fasten it to the floor. (The gap above is for air flow.)
8. In the front panel, drill a ⅞-inch hole into the smaller "vestibule."
9. Insert PVC pipe in exterior hole, flush with front panel. Add a drop of hot glue or caulk as needed.
10. Drill several ¼-inch holes in the floor for drainage and through the side panels (near the top) for ventilation. Fine screening can be stapled over the inside of the holes to help prevent access by uninvited guests such as ants.
11. Flip the box over onto the top panel with three of the sides 1 inch from the edges—more overhang on the front will provide a "porch roof"—and mark the outline of the box onto the top panel.
12. Cut the 1 × 1 into lengths—two pieces should be 15¼ inches and two other pieces should be 11¾ inches.

13. Fasten the 1 × 1s to the top of the box on your marks, so that the roof will slip over the box easily.

Optional: Add a Plexiglas panel to your box before installing the roof for safer inspections.

For the finishing touches, waterproof the roof with paint or roofing material, paint a flower at the entrance, and add legs to keep the nest box off the ground.

FURNISHINGS

You can fit corrugated cardboard on the floors of both "rooms" to soak up waste, and in the larger room—the brood chamber—place a handful of cotton batting or similar material for nest building. Some commercial bumblebee houses offer a small container of sweet syrup in the vestibule for the queen.

PLACEMENT OF THE BUMBLEBEE HOUSE

Your bumblebees won't want to be flooded, so choose a piece of higher ground to locate their new home. Place it on bricks or blocks to prevent water from rotting the bottom. They will enjoy the morning sun, so face the entrance east or southeast. Don't put it too close to a child's play area or on grass that you will be mowing; bumblebees are not aggressive, but they will sting if provoked.

Water, Water, Everywhere

Shelter? Check. Nectar and pollen sources? Check. But bumblebees, honeybees—*all* manner of bees, actually—also need a good source of water. That was news to me—until the day one of my neighbors stopped by and exclaimed, "Your bees are in our birdbath!" Mortified, I'd looked up from the black raspberry brambles I'd been tending to ask, "You mean *all* of them?" It was the height of summer, and the colony had been doing exceedingly well. Offering a string of apologies, I'd glanced nervously back at my fenced-in beehive, wondering what mischief the girls had gotten themselves into. But good news: It was only a handful of foraging workers taking a dip, and the old lady and her husband enjoyed watching them alight, walk along the birdbath's edge, and lower their tongues to suck up the precious resource. I knew it was lucky that my neighbors didn't mind the intrusion—lots of other people might have. Not one to press my luck, though, I had to do *something* to make sure that my bees didn't have to wander across the street every time they needed to load up on water.

To a honeybee, finding water is just as crucial as finding pollen and nectar. It's needed to help cool off the inside of the hive on stiflingly hot days, and nurse bees mix water with honey when preparing food for the hive's young. Some workers are so good at finding sources of water that it becomes their specialty; water foragers have been observed making as many as a hundred round-trips in a single day. The water they take in goes inside the same stomach that they usually use to store nectar, and once they're back inside the hive, they'll transfer their load to waiting bees. If water is badly needed, they'll dash off for more, and sometimes they'll encourage other foragers to fly out and fill up with them. In a multitude of tiny sips, these bees bring in nearly five gallons of water, collectively, each year.

Making a "Bee Bath"

To help keep any water-seeking bees out of other people's birdbaths and swimming pools, many beekeepers set out dedicated water sources for their bees, and making your own "bee bath" is just one more way you can help bees of every stripe without actually *keeping* them on purpose. Much smaller and lower to the ground than birdbaths, bee baths are scaled perfectly for bees, and they offer some protection from the purple martins, tufted titmice, and all the other feathered creatures that, given the chance, would just as soon eat one's honeybees as share a drink with them.

To make a bee bath, you'll need a shallow bowl or saucer; terra-cotta pot saucers work well. And to give drinking bees surer footing—and thereby prevent accidental drownings—find some pea gravel to sprinkle inside your saucer.[4]

Next, add water and plan to keep the bee bath topped off. (Because bees aren't picky about the quality of the water they collect, there's really no need to wash the bee bath between

4 In lieu of pea gravel, you can add tiny geodes, seashells, or even pretty glass marbles if you're so inclined.

fill-ups, and often, the mossier the saucers, the happier the bees seem.) Finally, to keep everyone happy, place these tiny oases away from heavily trafficked areas and near tall weeds or stands of flowers if possible.

The Honeybee Calendar

Eventually, your yard and gardens should be hopping with pollinators, and because they gravitate to many of the same nectar sources, you should spot many more butterflies and hummingbirds, too. Attracting all of them is one thing, but keeping them from flying away is another. As someone with literally thousands of "mouths" to feed, I realized I'd better have *something* in bloom all the time. So I grabbed an extra calendar I had lying around and started marking the days and weeks when different flowering bulbs, annuals, and perennials were blooming in my neighborhood. I call it my honeybee calendar, and I try to keep it up-to-date during the spring, summer, and fall months. The very first honeybee calendar I'd kept really helped me to spot holes in the landscape. Late spring and midsummer in my yard had brought an embarrassment of riches, but very early spring and late fall left a lot to be desired.

To better fill out my bloom time calendar, I planted flowering bulbs like crocus and alliums, and I sprang for some flowering annuals at the local farmers' market—both little luxuries I could never afford for myself. Ah, but it's for the bees, so that justifies the expense! I'm also justified in leaving a patch or two of weeds growing, since honeybees and other pollinating insects don't distinguish between the Queen Anne's lace gone wild and my carefully cultivated white yarrow. Both of those have large flower heads jammed with a multitude of much smaller flowers—something bees really love.

If you choose to start your own honeybee calendar, you'll probably be able to identify certain times in your area when you don't have as much in flower either. It doesn't take a green thumb to fill in any gaps you notice. There are loads of easy-to-grow plants that you can introduce to expand your bloom period, and here are just a few to get you thinking:

SPRING: Flowering bulbs like snowdrops and crocuses are your best bet for early and midspring flowers, and for a really natural look, you can scatter a few handfuls of these, and then plant them wherever they land. You might also try hyacinths, alliums, bluebells, and lily of the valley, and so you don't have to spend time weeding, you can intersperse a ground cover like creeping phlox between your bulbs.

SUMMER: This time of the year isn't usually much of a problem, and if you grow dill, fennel, purple coneflowers, lamb's ear, bee balm, salvia, borage, and stonecrop, you'll have summer—and in some cases, even early fall—covered.

FALL: Black-eyed Susans, joe-pye weed, and some of the other flowers that bloomed in summer can hold their own during much of the fall, and both asters and goldenrod are major food sources for honeybees at the end of the growing season.

Sounds like too much to handle? Many commercially available wildflower mixes contain some of the flowers honeybees like, so even if you simply shake a can of seeds over a prepared garden bed, water them well, and then forget all about them, the bees will get some good from what survives. Besides enjoying all the nearly year-round color that a good mix of flowers can provide, you'll also be helping any beekeepers who may keep their hives close by. The more natural nectar sources around, the more delicious honey the foraging worker bees will be able to make.

CHAPTER NINE

The Sweet Life

I'm pretty sure honey has magical powers, because, in one case at least, I've seen it transform a man. The old guy who lives behind me went from cantankerous to sugary-sweet just like that. All it took was a steady supply of my hand-labeled jars full of honey, and now he's a veritable Pooh Bear with a taste for the bees' annual crop.[1]

The funny thing is I'd never really planned to take any honey away from my bees to begin with. I figured they worked for it, and they should be allowed to keep it. But that was before I realized that if the bees are *too* productive, they will fill every nook and cranny with honey—after which time they unfailingly swarm, creating a truly unnerving neighborhood spec-

1 Truly I have grown to like him. The nicest thing anyone has ever had to say about my honey came straight from him. Here's the honest-to-goodness, direct quote: "I'mohnna tell ewe. That was the BEST honey ah have EVER had. Good body. More flavorful. You gotchoo sum GOOD bees now!"

tacle. To put a stop to that, I had to offer them more space. Since I didn't have enough empty supers and frames to add on to the existing hives, I would have to pull out a few of their filled frames, extract the honey they contained, and then put the now-empty honeycomb sections back into the hives for the bees to refill. I just wasn't so sure how.

Pawing through the spare beekeeping equipment I had on hand, I found a few plastic oval-shaped gizmos that I knew to be bee "escapes"—small, one-way "doors" that beekeepers insert in their hives a day or so before they remove full supers of honey. The oval-shaped insert was designed to fit inside the middle of an inner cover board, but I didn't know which side should face up and which should face down. It didn't look like there were any instructions for them either, so after asking around, I popped the bee escape into the cover board, marched out to the hive, and placed the board just under the super from which I would take several honey-filled frames. All

the beekeepers I asked said in about twenty-four hours the top super would be completely free of bees. Any bees that *had* been inside would eventually exit through the one-way bee escape, and even if they changed their minds and decided to pop back up there for some reason, they wouldn't be able to. That made me confident enough to stride out the next day sans bee smoker, jumpsuit, or veil. I lifted the top covers off my hive only to have hundreds of angry bees boil out and come after me. To this day I do not blame them. Because I installed the bee escape facing the wrong way, I'd unintentionally trapped much of the hive inside the already cramped top super.

There is a profusion of bee escapes that, in the right hands, are foolproof and effective. Aside from the plastic ones designed to fit inside inner cover boards, there are many other bee escapes in assorted shapes and sizes, and most of them function on the same one-way-door principle. But there are beekeepers who choose to drive their bees away via chemical rather than mechanical means. To clear any honey-filled supers to be commandeered, the beekeeper applies a small amount of very foul-smelling butyric anhydride to one side of a cloth-covered "fume" board.[2] This board is fitted, smelly side down, over the super to be removed, and within a few minutes, any bees previously occupying that level of the hive will have scurried out, leaving the beekeeper to cart the super off to his honey house. Still, other beekeepers—usually from the large commercial outfits—use gas-powered or electric bee "blowers," which quickly dislodge the bees with puffs of air.

2 Butyric anhydride smells like rancid butter mixed with vomit, and a little bit really packs a wallop. I once took the tiniest whiff for myself and nearly threw up.

But some people prefer less high-tech methods to separate honeybees from their honey. Using a wide, silky "bee brush," a beekeeper can lightly sweep bees off individual frames, one frame at a time. Once a frame is bee-free, it can be set off to the side in an empty super topped with a lid for safekeeping. As soon as the beekeeper has removed as many frames as he wishes, he'll spirit them away for honey extraction.

Honey Extraction 101

Although I bristle at the expression "robbing the bees," I admit that each time I carry a few frames, heavy with honey, inside to be extracted, I feel as if I've gotten away with real treasure. But it wasn't always this way. The first few times I harvested honey from my hives, I ended up with several bottles of yeasty, fermenting goo. The problem was I'd mixed a large amount of uncapped nectar in with what had been completely ripened, capped honey. Old-timers call that uncapped nectar "green" honey, and really, I would've been better off leaving any frames with large portions of it inside my hives, so that the bees could properly finish what they'd started. Because it still has a very high moisture content, green honey presents an agreeable enough environment for wild yeasts to flourish; fully ripened honey doesn't. To fix the problem, I had the bright idea to evaporate some of the water off by heating the freshly extracted honey in a large pot on my stovetop. Although I kept stirring the

---

The Guinness Book of World Records *reports a Greek beekeeper extracted the heaviest honeycomb to date; it weighed in at 22 pounds, 14 ounces.*

---

substandard fare, I only succeeded in making things worse. The honey got too hot, and in a flash, it darkened ominously and smelled like sweet, burning hair. Happily, by now, I have things figured out . . .

* * *

The average hobbyist's honey extractor is simple but elegant. Typically, it's a large, stainless steel drum that will hold just a few frames of honey at a time. The frames are suspended vertically in wire baskets connected to a hand crank at the top of the extractor. When that crank is turned, centrifugal force flings the honey out of the honeycomb's waxy cells and down the drum's smooth walls. Beekeepers with larger apiaries often use electric-powered extractors that can hold more than a hundred frames at once, but no matter how many frames one can process at a time, there's no getting around the fact that the wax-sealed honeycomb must first be uncapped, and that can be a very messy job.

More Tools of the Trade

There is a gadget for just about every part of beekeeping, and cell uncapping is no exception. The beekeeper prone to buying lots of fancy equipment (or in fairness, one who bottles a lot more honey than I do) will use an uncapping knife—that's a knife with a built-in heat element—to slice through the bees' waxy seals like so much butter. There, too, are lower-tech uncapping "forks" or "scratchers" that look a bit like those metal hair picks the 1970s set used to keep in their back pockets. Beekeepers drag the tines of these lightly over the capped wax to poke uniform holes over the honey-filled cells.

One lacking any of these "official" uncapping tools can seesaw a serrated kitchen knife just underneath the honeycomb's waxy surface, but it's difficult not to lose copious amounts of honey in the process. This honey, along with the loose wax bits, is set aside to be dealt with a little later, and after each frame has been uncapped on both sides, it's ready to go into the extractor.[3] Whether with electricity or elbow grease, the frames are spun, and the honey flies out and collects in the bottom of the extractor. The beekeeper might scrape the extractor walls to get every last drop, and when he's ready, he'll open the spigot and strain the honey through a few layers of cheesecloth. This removes any stray clumps of pollen and the occasional bit of honeybee wing or other debris, and finally, the honey is ready to bottle.

Tortured Honey Versus Tasty Honey

Many large honey-packing facilities won't bottle honey until it has been heated and filtered half to death. In order to "pasteurize" their honey, they bring it to very high temperatures, and sometimes they process it even further by shooting high-pressure streams of honey through a series of filters. If you've ever bought any at one of those big chain supermarkets or

3 Wax cappings can be wrapped in cheesecloth and squeezed in order to drain the last bit of honey clinging to them, and the leftover wax itself can enjoy a second life when made into soap, poured into candle molds, or should the need arise, fashioned into menacing voodoo dolls.

department stores, odds are you've already tasted honey that has been thoroughly tortured in these ways. The sad part of it is honey doesn't really *need* to be pasteurized at all, but the companies do it to destroy any lurking yeasts and to increase their honey's shelf life.[4] The pressurized filtration step removes air bubbles along with tiny bits of pollen and wax so that the final product looks perfectly clear—even if it has been robbed of some of its more subtle flavors.[5]

Truly, a spoonful of just-harvested, unpasteurized, unfiltered honey can taste more complex than a vintage port wine. There are at least 300 varieties of honey in the United States alone, and as with wine, honey's many flavors and colors are influenced by geographic location, rainfall, soil conditions, and time of harvest. Bruce Wolk, director of marketing for the National Honey Board, has noticed that chefs particularly prize it: "Especially at white tablecloth restaurants, chefs have been really pursuing the different flavors of honey and even the different aromas of honey. They value honey very much for its ability to flavor dishes differently."

Honey is never quite the same from year to year or season to season. What a beekeeper extracts in the spring is usually much clearer and lighter colored than honey from late summer, and while most people seem to like the taste of spring clover honey best, some swear the dark, really complex wildflower honeys reign supreme. Foodies with highly sophisticated

4 High temperatures during the pasteurization process destroy the seed crystals that are naturally present in raw honey. Eliminate them and the honey isn't as apt to granulate and ferment.

5 One thing honey doesn't lose? The *Clostridium botulinum* spores, which, although rare, can be present in both pasteurized and raw honeys. They don't pose much of a threat to adults, but infants under one year old shouldn't be fed honey.

palates seek out honey made solely from individual flower types, and although it may seem difficult, isolating specific honey varieties is actually pretty easy. Tracy Hunter, a successful Indiana-based beekeeper, sells a dozen different honey varieties, including black locust, alfalfa, apple blossom, blueberry, and even lavender. "A beekeeper has to be in tune with what's in bloom and what his bees are working and then be able to remove that before it is mixed with any other honey in the hive," he says. The fact that bees will work just one flower type at a time makes isolating different honey flavors a little easier. Say a worker is attracted to a dandelion in the yard. She will devote her full attention to that specific flower variety—and furthermore, to that individual flower—until that dandelion is all dried up. "A majority of the bees are working one type of flower, so they won't mix the honey in their bodies, and when you look in the hive, you'll see that each type of honey is put together. You'll see different colors of honey in different sections of honeycomb," Hunter says.

Those colors can run the gamut from nearly white to dark chocolate brown, but somewhere along the way, it was decided that consumers were willing to pay more for the lighter honey than the darker stuff. That complicated things for commercial beekeepers and honey packagers, who didn't always agree on what counted as high-priced "light" honey and what should sell as the lower-rated "dark." Official standards for honey hues are now in place, with honey ranging from "water white," "white," and "extra light amber" to "light amber," "amber," and "dark amber." To see just how their honey stacks up, some beekeepers eyeball color swatches akin to the paint sample strips one might pore over in a home improvement store or, more aptly, the blonds, reds, and rich browns on display at the beauty parlor. But as you might guess, there are people for whom color swatches simply won't do. To determine whether this season's crop is "water white" or just a run-of-the-mill "amber," they use digital honey analyzers to precisely measure the amount of light absorbed by each honey sample, and then they compare their readings to those of widely accepted honey color standards.

"Mad" Honey

Color and flavor aren't the only qualities that can set one kind of honey apart from another. Bees foraging almost exclusively on certain types of rhododendrons—especially *Rhododendron ponticum*, which grows in the mountainous areas of Turkey—make a truly memorable honey. Sometimes called "mad honey," once ingested it can cause dizziness, nausea and vomiting, irregular heartbeat, and more. The culprit? Dangerous grayanotoxin contained in the rhododendrons' nectar. Sometimes known as rhodotoxin, grayanotoxin is a naturally occurring toxin found in certain types of rhododendron plants as well as mountain laurel and sheep laurel. Just three and a half tablespoons is enough to make one sick, but happily, time is usually an adequate antidote. The symptoms of grayanotoxin poisoning disappear in about twenty-four hours for most people, but in severe cases, doctors sometimes will administer treatments to mitigate the more serious side effects, including loss of coordination and convulsions.[6]

6 After eating large amounts of toxic leaves, some goats have died.

In his famous work *Anabis*, Xenophon describes the Greek army's run-in with grayanotoxin:

> *The effect upon the soldiers who tasted the combs was, that they all went for the nonce quite off their heads, and suffered from vomiting and diarrhoea, with a total inability to stand steady on their legs. A small dose produced a condition not unlike violent drunkenness, a large one an attack very like a fit of madness, and some dropped down, apparently at death's door. So they lay, hundreds of them, as if there had been a great defeat, a prey to the cruellest despondency. But the next day, none had died; and almost at the same hour of the day at which they had eaten they recovered their senses, and on the third or fourth day got on their legs again like convalescents after a severe course of medical treatment.*

Scholars believe mad honey was even left intentionally for others to find. In *The World History of Beekeeping and Honey Hunting*, author Eva Crane points to Volume 12 of Strabo's *Geographica* for proof:

> *Strabo . . . described the military use of toxic honey from the same region in the Third Mithridatic War, probably in 65 BC, and it is likely that this honey . . . came from* R. ponticum. *The Heptacometae, inhabitants of Pontus, placed on the road by which Pompey's soldiers would pass through the mountains "vessels filled with maddening honey, which is procured from the branches of trees. The men who had tasted the honey were attacked and easily*

despatched." About 1,200 men were probably killed on this occasion.

Actually getting one's hands on enough of that nectar to do the job wouldn't be easy, so even if you did have a couple of rhododendrons in your yard, you needn't be worried.

Honey and the "Halo Effect"

To hear some people talk, you'd think honey could cure cancer, save the whales, and bring about world peace. Oodles of claims have been made about the benefits of honey, and they've been repeated so many times that we've come to accept them as fact. But not so fast. Yes, there are lots of things about honey that are technically true. For one, honey contains protein, antioxidants, amino acids, vitamins, and minerals, but they're present in such tiny amounts that you'd have to eat dozens of cases of honey each month to make it really count. As for all the people who believe eating local honey, which is chock-full of pollen from area flora, will help to diminish the effects of certain allergies, so far at least, scientific studies haven't been able to prove unequivocally that there is a connection between eating honey and easing one's sneezing and sniffling. Same goes for honey's

There are sixty-four calories per tablespoon of honey.

antibacterial and antimicrobial properties. "There have been a lot of attributes ascribed to honey that really don't stand up under the scrutiny of scientific investigation," admits Bruce Wolk of the National Honey Board. "Most of the research that's been presented in terms of the medicinal benefits of honey have largely come from overseas and really have not gone under the scrutiny of a legitimate, double-blind study by a medical school or research institution in [the United States]."

That's not to say honey doesn't offer *any* benefits. A small study conducted by researchers at Penn State College of Medicine revealed that a dose of honey given to kids with upper respiratory tract infections outperformed doses of cough medicine.

There are a few other areas in which honey really shines. Wolk says honey's the only ecofriendly sweetener around: "Virtually every other sweetener has a net-negative effect on the environment. It has a large carbon footprint. Honey doesn't." Scandinavian researchers analyzed the different energy costs associated with growing, harvesting, processing, storing, and transporting different types of sweeteners including honey as well as sugars derived from sugarcane, beets, and corn. I think what they found finally justifies putting a halo over honey. While sugarcane, beets, and corn require irrigation, fertilizers, pesticides—and in some cases, a ton of backbreaking human labor in Third World countries—honey does not. It's simply one of the by-products honeybees make, and as they gather assorted floral nectars, they offer their invaluable pollination services to boot. Besides that, honey doesn't need to be shipped in from Cuba, the Dominican Republic, or even farther reaches like sugar does, so all in all, honey's a pretty sweet deal.

* * *

Although decidedly more mundane, another of honey's benefits is its humectant property. Because it retains moisture so well, you'll sometimes find it in soaps, lip balms, and creams, and bakers like to use honey for extra-moist breads, cakes, and brownies.

Caveat Emptor

Want to start reaping the benefits of honey yourself? Before you put that cute plastic honey bear in your cart, you'd better read the label carefully. The U.S. Food and Drug Administration has no labeling standards for the commodity, and that has allowed a lot of honey pretenders to sneak in. Wolk explains, "When you see something that says 'honey blend,' for example, most of us take that to mean, well, it might be a combination of tupelo, orange blossom, and clover, but that's not what those labels imply at all. Those labels imply that they're taking honey and they're cutting it—maybe 50 percent, maybe more—with another filler that's much cheaper." Sometimes manufacturers will add high fructose corn syrup or maltose syrup and honey "flavorings" to cut production costs, and if you read the fine print, you'll have a better idea of just what you're spooning into your tea or spreading onto your toast.

To make matters worse, there are loads of products with the word "honey" in their name that don't contain much of the real thing. The next time you're at the grocery store, take a look at some of the labels. If honey is listed right above the preservatives, you'll know there's virtually none in there, but if it's listed closer to the top, you can be assured you're getting what you paid for.

Yummy Honey Recipes

Once you're pretty sure you've got the real thing, there are all sorts of uses for honey you can try. For instance, I once helped a friend make several batches of honey wine. Producing five gallons of an extra-hard, extra-sweet mead required eighteen pounds of raw honey, and although the result was decidedly ambrosial, it seemed almost sinful to use so much honey. So, here are a few recipes that don't require quite as much of the sweet stuff.

Fat-Free Honey Herb Dressing

Makes ½ cup

¼ cup white wine vinegar
¼ cup honey
2 tablespoons chopped fresh basil or mint
1 tablespoon minced green onion
Salt and pepper, to taste

In a small bowl, combine all the ingredients. Mix well.

Curried Honey Sweet Potato Soup

Makes 8 cups

1 tablespoon olive oil
1 onion, diced
4 medium-sized cloves garlic, peeled
6 cups (48 ounces) chicken or vegetable stock
1 pound sweet potatoes, peeled and cut into chunks
1 medium russet potato, peeled and cut into chunks
2 teaspoons salt
6 tablespoons orange blossom honey, divided
1 medium red bell pepper, seeded and diced
2 to 3 teaspoons curry powder
½ teaspoon pepper
½ teaspoon ground ginger
¼ cup chopped fresh cilantro, optional

Heat the oil over medium-high heat in a soup pot. Add the onion and sauté until translucent, 2 to 3 minutes. Add the garlic and sauté 1 minute. Add the stock, potatoes, and salt. Cover and simmer until the potatoes are tender, about 15 minutes.

Puree the mixture in batches, put the soup back over low heat, and add 5 tablespoons of the honey, bell pepper, curry powder, pepper, and ginger. Bring to a simmer, taste, and adjust seasonings.

Microwave remaining 1 tablespoon of honey for 5 seconds on high. Serve the soup drizzled with a little warm honey and sprinkled with chopped cilantro, if desired. Serves 4 to 6.

Fire House Energy Bars

Makes 16 servings

½ cup butter or margarine, melted
2 tablespoons honey
⅔ cup walnuts, sliced or diced
2 eggs, beaten
2 cups granola cereal
1 teaspoon vanilla

Preheat oven to 350°F. Place all ingredients in a large mixing bowl. Blend well. Pat into a greased 8-inch square baking dish. Bake for 18 to 20 minutes or until lightly browned. Cool and cut into 16 bars.

Baklava

3 cups finely chopped walnuts
2 teaspoons ground cinnamon
½ teaspoon ground nutmeg
Ground cloves
1½ cups clarified butter, divided (see below)
½ cup honey
1 package (16 ounces) filo pastry sheets
Honey Syrup (see page 149)

Preheat oven to 325°F. Combine the walnuts and spices in a medium bowl. Reserve ½ cup clarified butter for brushing the top and bottom layers; stir the honey into the remaining 1 cup butter. Brush the bottom of a 13 × 9 × 2-inch baking pan with clarified butter. Cut the filo sheets in half crosswise; trim to 13 × 9-inch rectangles. Cover the filo with waxed paper and a damp towel to keep it from drying out.

Line the pan with 10 sheets of filo, brushing each with clarified butter; sprinkle with ⅓ cup walnut mixture. Place 2 sheets filo on top of the walnut layer, brushing each with honey-butter mixture. Sprinkle with ⅓ cup walnut mixture. Repeat, layering 2 sheets filo, brushing each with honey-butter mixture and sprinkling with ⅓ cup walnut mixture until all the nut mixture is used. Top with remaining filo sheets, brushing each with clarified butter. With a sharp knife, cut the baklava into diamond-shaped pieces, carefully cutting through all layers.

Bake for 45 minutes. Reduce heat to 275°F and bake 20 minutes more. Remove from oven; while still hot, carefully spoon cool Honey Syrup over the entire surface.

To clarify butter: Cut 1 pound butter into pieces and melt in a medium saucepan over medium heat. Skim off the foam; strain the clear yellow liquid into a bowl, leaving the cloudy residue in the bottom of the pan.

Honey Syrup

1 cup honey
¾ cup water
½ teaspoon grated lemon peel
3 whole cloves

1 cinnamon stick, 3 inches long
1½ teaspoons lemon juice

Combine the honey, water, lemon peel, cloves, and cinnamon stick in a small saucepan over medium heat. Bring to a boil. Reduce heat to low and simmer 20 minutes. Add the lemon juice; simmer 5 minutes more. Remove from heat, and cool. Remove the cloves and cinnamon stick before using.

Minding Your Beeswax

Having bees means having a lot of beeswax, too. My honeycomb cappings alone quickly pile up. I keep them in my freezer until I have enough to bother with, and once I do, I place the large chunks, which are still sticky from the residual honey, into the legs of a pair of old pantyhose. I tie knots above the wax and place the whole thing into a pot of water, slowly heating the pot until the wax in the hose begins to melt. The pure beeswax floats out of the nylons and up to the surface, forming a skim. Once I think I've gotten most of the wax, I remove the pantyhose, which now contain what beekeepers call the "slumgum"—a hodgepodge of dark brown clumps of propolis, pollen, bits of old cocoons, and any other leftovers. As the water in the large pot cools, the wax at the top hardens into a fragrant, cream-colored disk. This is rendered beeswax, and it can be used to make candles, soaps, and lotions. As with my early attempts to extract honey, my first crack at candle making was a spectacular failure. (If only I'd realized that *before* I'd given them to friends and family as gifts.)

I suppose I should've known something wasn't quite right just by looking at them. Rather than a buttery yellow, my candles were greenish, tinged like the yolk of an egg that's been boiled for far too long. Moreover, their texture was sandy, not smooth. I had been so eager to enjoy the warm glow of handmade candlelight that I hadn't bothered to render the beeswax completely. My candles were roughly equal parts slumgum and beeswax, and when lit, they hissed and popped strangely. Sometimes the wicks became dislodged and fell right out of the candles' centers, and of those that retained their wicks, many spontaneously extinguished themselves, as if out of shame for having been so shabbily crafted. Nevertheless, I tried to burn a few of them, I rubbed one along the edge of a dresser drawer to make it open more smoothly, and at some point accepting defeat, I threw the rest of the votives back in the freezer with the next batch of raw wax cappings.

* * *

I've gotten better at rendering my wax, and at some point, I decided I should try making my own beeswax soap. As you might imagine, that turned out about as well as my first candles had. I'd searched hardware stores for lye flakes, and finally found some, billed as a supercaustic drain cleaner and sporting a bright red skull and crossbones. I'd need olive and palm oils also, but coming up empty on the palm oil, I assumed I could substitute the same amount of coconut oil instead.[7] That was the first in a series of blunders that would contribute to the foulest "soap" ever.

7 Palm oil isn't as hard to find as I originally thought. Many of the solid vegetable shortenings sold at health food stores are made entirely from palm oil.

My daredevil boyfriend had come over to help, and immediately preoccupied himself with the lye. Seeking some connection with one of his literary heroes—Tyler Durden of Chuck Palahniuk's fantastic *Fight Club*—he gave the back of his hand a chaste little kiss and sprinkled a few lye flakes over the top. Because he hadn't gotten his hand wet, nothing special happened, but soon we would both develop a new regard for the alkali's power. My attention had wandered somewhat, and the small pot of lye water I'd been tending overflowed impressively.

Pressing on, I mixed the fats with the lye, paying no attention to the temperature of either, added a few drops of vanilla oil, poured the finished product into a plastic pan, and waited for our abomination to cure. We noticed a few small pools of water standing on the surface of the soap, and inexplicably, my boyfriend dipped the tip of his tongue into one of them.[8] Just as his flesh started to burn, I recognized the puddles for what they were—diminutive lakes of leftover lye water. (By the way, should you ever have an unfortunate Tyler Durden moment of your own, you can neutralize the lye by rinsing with vinegar.) Despite that extra lye, we allowed the soap a couple of weeks to cure, and when it finally was time, we excitedly cut the contents of the pan into small bars, and took them to our respective bathrooms to try out. The soap hardly lathered at all, smelled like cake mix, and left me covered with a slightly gritty, oily residue. (Sadly, I was forced to wash the homemade soap off with the store-bought kind.)

But working with beeswax need not be disastrous. You can

8 Curious men. Why *do* you do the things you do?

often buy perfectly rendered bars of it from local beekeepers, candlemaking supply houses, or craft stores. And if you plan to make your own soap, you can find most of what you need at health food stores. Here are a few beeswax projects to try.

Can-Do Candles

Newspapers
Candle mold
Vegetable oil spray
Candle wicks
Modeling clay or Plasti-Tak
Old long-sleeved shirt (optional)
Oven mitts
Beeswax
Old coffeepot or pan solely for melting beeswax
Small knife (optional)

To make decent candles, you must start with properly rendered beeswax. That means no honeybee antennae, legs, or pollen allowed. You'll also need a mold, which you can get at most craft supply stores. The mold I use is made of plastic and makes six votive candles that are an inch and a half in diameter. And then comes the candlewick. While you can buy whole spools of candlewick, I recommend

getting precut, measured wicks with metal bottoms, since you're just starting out. The package label should tell you for which candle diameters the wicks are suited, so once you've chosen your mold, you can find compatible wicks. (The wicks I use are suitable for use in candles that are 1 to 2 inches in diameter.)

Spread a few thick layers of newspaper down to start, and make sure the inside of your candle mold is clean and dry. If you see any debris, wipe it away; otherwise, it will be stuck on the outside of your finished candles. Next, spray the inside of your mold with cooking oil, so your candles will come out easily once it's time. At this point, I stick the wick ends, one through each hole, through what will be the tops of my finished candles. As you do this, make sure each wick stays straight and that the metal bottom attached to it is flush with what will be the bottom of your candle. Turn your attention back to the top of your candle, bending the wick slightly to help it stay in place. For the next step I use Plasti-Tak—that sticky putty that I used in high school to hang up my Pink Floyd posters—but you could just as easily use modeling clay. You'll want to secure the tops of your wicks with this on the outside of your candle mold. That way the wicks won't slip out of place, causing beeswax to stream out of the little wick holes in your mold. But if you do have a lavalike beeswax flow, the thick newspapers are there to catch any spills or leaks.

Now to the fun part. An old long-sleeved shirt and oven mitts are nice to have in case you accidentally splash yourself with hot wax, so put those on and carefully melt the beeswax in a pan. (Because beeswax is a real pain to clean up, it's best to have one pan solely devoted to melting it, but even an old metal coffeepot with a nice pouring spout will work.) As the wax starts to melt, keep the pan moving—I don't even let mine touch the burner on my gas stove—because beeswax is rather flammable stuff. Let the hot wax cool just a little in its container, and then slowly pour it into your mold until it reaches what is the lip of the mold and the bottom of your candle. (Warning: If you didn't seal the wicks properly in place and cover the wick holes completely with modeling clay or Plasti-Tak, some of the wax will find its way out of the mold and onto the newspaper.)

As the wax cools and hardens, you may notice that there is now some extra room at the top of your mold. Melt more beeswax and top off any candles that came out low on wax. Once the beeswax has cooled and hardened completely, you can turn the mold over, and gently cut away any extra wax that managed to seep through the wick holes. Set any of this extra wax aside to be rerendered later. Almost finished! Pop your candles out of the molds, and if you like, you can gently trim any excess wax from around the wick. Please don't feel bad if your candles look awful. You'll get the hang of it with practice, and anyway, the beauty of beeswax is that you can always remelt it and try, try again.

DIY "Death Mask"

Oh, but aren't candles a little humdrum? If you've got a taste for the macabre, I wholeheartedly recommend the do-it-yourself death mask instead. Some of the most famous ones—like those of King Louis XVI and Marie Antoinette—were cast in beeswax by Madame Tussaud, skilled in the art of figure sculpting and easily the indubitable queen of beeswax.

According to Kate Berridge, the author of *Madame Tussaud: A Life in Wax*, Tussaud was also the queen of spin:

> *Having presented herself as a victim and survivor of the French Revolution, Madame Tussaud remains for all time suspended in people's imagination as a young woman with a guillotine-fresh head in her lap. The image of an innocent woman in a bloody apron being forced to make death heads to save her own neck elicited both sympathy and curiosity in her public.*

Even now, Madame Tussaud's name—and her stories—live on. Tour any of her thriving waxworks museums, and you're sure to hear that she made those death masks of Louis XVI, Marie Antoinette, and their contemporaries under duress. It's more likely, though, that she actively sought to profit from the public's ghoulish fascination with the carnage of the day.

Fast-forward to the present, and you'll find less gore and more star power in the exhibits bearing her name. There are waxen replicas of Britney Spears, Shaquille O'Neil, George W. Bush, and other modern notables she would never meet. Some things don't change, though: All of the figures in Madame Tussaud's wax museums are still made with beeswax. Since today's wax studio artists don't necessarily have the actual subjects at their disposal, they must work from photos and measurements to sculpt clay models. They then make plaster casts of the models, and once those have hardened completely, the casts are filled with hot beeswax tinted with color to match the real people's various skin tones. For the last step, glass eyes are installed, human hair is inserted—one hair at a time—into the celebrity scalps, and then they are carefully painted, to incredibly lifelike effect.

Since you're only making a death mask and not a full, waxy homage to yourself, you won't have to go to nearly all that trouble. Better still, you don't even have to be dead.

MATERIALS

- Lots of beeswax
- Metal pot
- Swim cap or plastic wrap
- Petroleum jelly
- Alginate/nontoxic life casting mold mix
- Water
- 2 large mixing bowls
- Whisk or electric beaters
- Extra-large drinking straws
- Scissors
- Spoon
- Old towel or plastic tarp
- Pad of paper and pen
- Sand
- A friend you trust not to kill you

Before you subject your face to any life casting product, make sure you won't have an allergic reaction to it by mixing a small batch to test on your foot or hand. Once you're sure it's okay, you can put on the swim cap or plastic wrap to cover your hairline while your friend works with the life casting mold mix. The one I like calls for equal parts of life casting mold mix powder and water, but yours might be different. Your pal should stir that vigorously to break up the clumps. Meanwhile, smooth the petroleum jelly along your hairline if any of it is still showing and over your eyebrows. When it's time to close your eyes, you'll also carefully jelly your

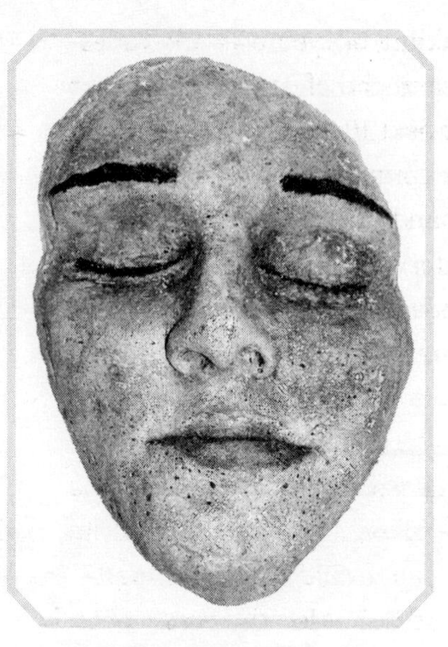

Made entirely out of beeswax, this "death mask" is eerily lifelike and luminous.

eyelashes.[9] Snip one extra-large drinking straw into four sections, and insert two of the sections into each nostril. You'd better be sure that the straws fit comfortably and that they won't slip out, because, for a good 10 to 20 minutes, they will be your only source of air.[10] At this point, lie down on the towel or tarp, which will catch any drips, and keep your eyes and mouth shut tightly from now until you're free of the molding compound. By now your friend should have the mold mix ready, and he or she can carefully drizzle the liquid life casting mix onto your nose and around each nostril with the spoon, taking pains to leave your breathing straws untouched.[11] After that delicate operation, pour most of the rest of the mix onto the face.[12] Keep your face frozen—

9 If you have a mustache or beard you'd like to keep, cover those with a layer of petroleum jelly as well.

10 If you have a cold or sinus infection that makes breathing through your nose difficult, this is obviously not the project for you!

11 Mixed properly, it should have the consistency of pancake batter, but because the life casting compound sets up fairly quickly, you don't have much time to obsess over every little lump.

12 This is about the time some face models get extremely nervous. Try to stay calm, breathe through your straws, and just think how cool your death mask is going to look.

no smiling or frowning, no crinkling of eyebrows. Charades-like hand gestures and that pen and pad of paper will be your means of communication for the next 10 minutes or so.[13]

Once the mold material has completely hardened into a rubbery shell, put both of your hands against your face, while slowly sitting upright. Now lean forward, and gravity will help the mask come off into your hands. Gently pull the straws out of the mold and then turn it around in order to dab a little extra mix into each nostril. While that addition hardens, fill the other large mixing bowl with sand, and then place the finished mold with the negative image of your face up. Use the sand to thoroughly support the mold, especially around its edges. If you peer into the mold, you should see great detail—right down to every line, pore, and hair. Melt the beeswax on the stove, pour it into the mold, and let it sit overnight. When the wax has cooled and solidified, you can remove the beeswax mask. Unfortunately, you may have to tear the delicate mold to get to it, but if you do want to use the same mold repeatedly, have your friend apply plaster-impregnated gauze strips to the outside of the mold form while it's still on your face. This will make your mold much sturdier, but waiting for the plaster bandages to harden up can take another 10 minutes or more. After that, follow the same steps to create your creepy wax visage.

You may have to do a little fine-tuning where the straws used to be. I used a small knife and wooden carving tools to clean up my waxy nostrils. As a finishing touch, I painted the

13 I did a lot of pointing to my watch and then turning one hand palm up, to ask, *Is it almost time to get this mess off my face?*

wax with a very watery acrylic paint mix to match my own skin tone, and I added a little color to my lips, eyelashes, and eyebrows. My death mask looks real enough that, if I squint a bit, I can envision it resting in a satin-lined casket with a smattering of my relatives weeping nearby.

Susan's "Respect the Lye!" Beeswax Soap[14]

3 ounces grated beeswax[15]
5 ounces coconut oil
4 ounces olive oil
4 ounces palm oil
2.12 ounces lye
5 ounces distilled water
Goggles
Rubber gloves
2 wooden spoons
Plastic measuring spoons
Plastic spatula
Metal pot
2 glass candy thermometers
Ceramic or glass mixing bowls
Glass measuring cup
Kitchen scale that measures in ounces
Petroleum jelly
Plastic or glass soap mold[16]
Vinegar

First, slather petroleum jelly onto the bottom and sides of the inside of your mold. (I use one of those plastic trays that fits inside a cashbox, but there are all kinds of mold possibili-

14 Many thanks go to the folks at Majestic Mountain Sage in Logan, Utah. They have an online lye calculator that made coming up with this recipe painless, and happily, I no longer make pans full of unusable goo.

15 I like to grate mine with a hand grater I use just for soap and candlemaking, because finely grated beeswax melts more quickly and evenly into the other fats.

16 This recipe makes a pound of soap and will fit perfectly into a 9×3×2-inch pan.

ties.) Set your greased mold aside for now.

Next, with the precision of a chemist, weigh all your ingredients except for the lye. Seriously, be very exacting here. Put the oils and beeswax together in your metal pot, and set that aside. Fill the glass measuring cup with distilled water. (Again, precision matters.)

In soap making, accurate measurements are a must.

On the scale, make sure the weighing bin—or whatever plastic container you're using to contain the lye flakes or crystals as you measure—is clean and completely dry. Like fire, lye is useful, but its power must be respected, so keep the lye-neutralizing vinegar close at hand in case of accidents. Now, put on your goggles and rubber gloves. Using a small, plastic spoon, measure the 2.12 ounces of lye, then go outdoors with the water-filled measuring cup, measured lye, wooden spoon, and one of the glass candy thermometers for the next step.[17] Be sure to find a safe place away from kids and pets.

Now it's time to mix the lye. Slowly pour the measured lye into the water—never the other way around, because it can splash and burn! Stir gently, and the lye water will heat up and

17 Lye produces dangerous, caustic fumes, and so that I don't end up breathing them, I've learned to mix my lye solution outside.

become transparent. Stick the thermometer in, and you'll be amazed at how hot this stuff gets. Leave it to cool. The target temperature will be 120°F for both the lye solution and your mixture of beeswax and oils, and getting both to exactly 120°F at the same time can be tricky.

Go back inside to heat the fats on the stove top, slowly mixing with the other wooden spoon. To melt everything together, the temperature of the wax and oils will go well above 120°F. When all has melted, gently pour the fatty liquid into a large glass or ceramic bowl, and stick in the other glass candy thermometer. At this point I'll take this bowl outside and put it next to the lye water to compare the temperatures on each. When the fat and the lye fall to 120°F, then it's time to pour the lye into the bowl of mixed fats. Stir this with the wooden spoon you originally used for mixing the beeswax and oils. As the lye binds with the fats, the mixture will become thick

Homemade bars of beeswax soap.

and creamy like vanilla pudding. It takes me anywhere from 10 to 20 minutes of mixing before the consistency is right. When you can drizzle the mixture off your spoon and into the bowl with the drizzled lines resting on the top for a second or so, you've reached the point soap makers call "trace."

Utah is known as "the beehive state," and the honeybee is the official state insect of Arkansas, Georgia, Kansas, Louisiana, Maine, Mississippi, Missouri, Nebraska, New Jersey, North Carolina, Oklahoma, South Dakota, Tennessee, Utah, Vermont, Wisconsin, and West Virginia.

At trace, pour the soap-to-be into your mold and smooth the top with a spatula as needed. Lightly press a sheet of plastic wrap onto the top of the soap, and let it "cure" or harden in the mold for a few days. You're almost there. . . . Turn the soap out of the mold and let it cure awhile more before cutting it into bars. You should end up with one pound of pleasing yellow soap. I cut mine into six small bars.

Nancy's Lip Balm

1 cup grated beeswax
14 ounces coconut oil
5 tablespoons honey
5 tablespoons pure vanilla extract

Heat the wax in a saucepan over low heat to 150°F. In a separate saucepan, heat the oil to the same temperature. When both are heated to the proper temperature, add the coconut oil to the beeswax, remove the pan from heat, and stir steadily until well blended. Then add the honey and the vanilla extract and continue to stir until well blended. Pour into tubes or tubs, allow to cool overnight, and then cap the containers and store at room temperature, out of direct sunlight.

Other Bee Goods

For every person who believes honey is a miracle food, there are several more who feel the same way about bee pollen, propolis, and royal jelly. It was Ronald Reagan who helped spur the pollen boom of the 1980s. "President Reagan ate two things that were a little unusual. If you went to the White House, he always had a jar of jelly beans on the desk, and he used to keep pollen bars in the refrigerator," says John Ambrose, a professor of entomology at North Carolina State University. "They were sort of the precursor to the nutrient bars we have now, and every day, supposedly, he ate two pollen bars." Pollen is still thought to increase stamina, eliminate allergies, promote weight loss, lower cholesterol, and reduce the signs of aging. The list of pollen's presumed benefits goes on and on and on, but Ambrose adds, "The idea of pollen being nature's perfect food, I always tell my classes, 'It is—but only if you're a bee!' So much of it is not digestible, and then there's the cost factor and everything else."

Pollen is pricey because it takes a lot of the tiny grains to pack into "energy" bars or fill up the bottles in your local health food store. Beekeepers collect it by attaching pollen traps to hive entrances. When pollen-loaded workers return to the hive, they have to pass through the trap on their way inside, but because the trap's entry points are extremely narrow, the

Assorted pollen grains from a sunflower, morning glory, hollyhock, lily, primrose, and castor bean viewed under an electron microscope.
Dartmouth Electron Microscope Facility, Dartmouth College

bees must scrape most of the pollen from their leg baskets in order to make it past the trap. The pollen they scrape off then drops into an isolated tray for the beekeeper to collect later. It seems a shame that a fully loaded worker bee should have to leave most of her "groceries" outside in this way, and as if that weren't discouraging enough, sometimes bees lose legs or wings trying to get through the contraptions. Also, colonies with attached pollen traps end up working even harder. "Bees will compensate for pollen traps, so if you put a trap on, the bees will actually collect more pollen to make up for the deficiency," Ambrose says.[18]

18 There is at least one good reason to use pollen traps. When pollen is plentiful, beekeepers sometimes will collect it with the intention of returning the stored pollen to their bees in the early spring. This helps bee colonies to more quickly build their populations—even if pollen sources outside are still a bit scarce.

All that and researchers have yet to definitively prove that pollen does people any real good. In fact, ingesting pollen might even be harmful. Consider this: Around the same time that Reagan was polishing off those pollen bars, many orchardists were relying on "microencapsulated" pesticides such as PennCap-M to control insect pests over long periods of time.[19] The insecticide was contained in tiny capsules to slow down its decomposition. "From a growers standpoint that was good, because they could put it out there and it would kill the pests longer," Ambrose says. Not wanting to harm the valuable honeybee pollinators, the orchard growers knew better than to apply the pesticide while their fruit trees were in bloom. They waited until the orchard bloom had ended to spray, but Ambrose points out that area bee colonies were dying anyway. "Bee colonies were dying in the late winter and early spring. We would take samples and find the pesticides that you would find in PennCap-M. Nobody was applying that then, because there was nothing growing. So, it was sort of a mystery."

Soon enough, the mystery was solved. The pesticide microcapsules were about the same size as pollen, and honeybees foraging on clover, dandelions, and other flowers directly under treated trees were picking up both the pesticide and grains of pollen and carrying them back to the hive either to feed larvae currently being raised or to put into long-term storage. Bees that were fed the pesticide-tainted pollen right away would die right away, and those larvae fed the bad pollen in the spring—along with the nurse bees feeding them—died

19 The heavy-duty pesticide originated from World War II nerve gas research.

in the spring. The use of PennCap-M was largely discontinued, but highly toxic pesticides and microencapsulated products still abound, and I imagine some of them do make their way into honeybees' pollen stores.

For the Good of Bees

To raise awareness of the dangers posed to insects and wildlife, the Environmental Protection Agency separates pesticides into three groups and requires manufacturers to explicitly label their products. "If a pesticide is highly toxic to bees—that is, it's a Category 1—the Environmental Protection Agency requires it to have a warning label saying, 'This product is toxic to bees, mammals, aquatic life, birds,' that sort of thing," Ambrose says. You might think most Category 1 pesticides are available only to big agricultural businesses,

but you'd be wrong. Actually, there are plenty of them on the shelves of most home improvement stores, and if you look, you might even find some in your own garden shed.

It can be hard to break the pesticide habit—especially when the Japanese beetles are ravaging your rosebushes for the third consecutive year—but doing without them really is best for area honeybees. In my own garden, I handpick the beetles mating on my asparagus plants and spray any aphids I see with a solution of soapy water. Planting some of the same plants honeybees like also helps to attract beneficial insects such as lacewings, parasitic wasps, and ladybugs, which, in turn, prey on my insect pests.

Still, if you think you *must* rely on a pesticide, reach for a Category 3 or Category 2 first, and look for products that break down quickly. Most important of all, apply the pesticide at a time when honeybees and wild pollinators aren't out foraging. "Up North, the best time would be in the evening, and then, if you have a pesticide that has a half-life of twelve hours, if you apply it at eight o'clock at night, . . . by eight o'clock the next morning, it's going to have lost half of its toxicity, so even if bees do come into contact with it, it's going to be safer," Ambrose notes. And for people living in the South, he recommends applying their pesticides in the late afternoon: "What happens, as you go further south, is you tend to have higher temperatures, and the plants stop producing nectar—or the nectar is all evaporated—by the time you apply it."

* * *

Collected a lot like pollen, propolis is another honeybee commodity with supposed health benefits, but by now, I'm guessing you know what I think of that. . . . Collecting royal jelly is even more complicated. Beekeepers must open up the queen cells and scoop out the pearly substance with a toothpick. Some large beekeeping operations use tiny vacuums about the size of a pencil to suck out the royal gloop. It takes well over 100 queen cells to come up with just a couple of tablespoons' worth. And for what? So that cosmeticians can entice aging ladies to plunk down eighty bucks a pop for tiny tubes of Gelée Royale Wrinkle Defying Serum? Well, crinkles and all, I plan to grow old with a little dignity. While I'll always enjoy a spoonful of honey and my beeswax soap, I'll be leaving the royal jelly for the *real* queens.

FINAL NOTE

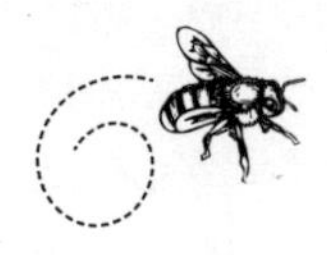

Who Needs Chickens Anyway?

I just set my bees up in a fancy new hive. It has a pitched copper roof and a sloped landing board at the base of the hive entrance, so that workers coming in with full honey stomachs can make their way inside with a little more decorum than usual. I painted the supers a pale baby blue—one of the bees' favorite colors—and left the roof untouched, so that it will gradually develop its inevitable turquoise patina. Things are looking pretty good on the outside, but after a recent hive inspection, I'm not so sure about the bees within. Characteristically, I never spotted the queen, nor did I see any eggs or larvae, but maybe I didn't look as carefully as I should have. There were, at least, lots of capped brood cells, with some new worker bees chewing their way out, completely oblivious to me. And foraging workers were still coming in with pollen, just as others were heading out to seek their respective fortunes. Not bad, considering they'd made it through the winter with only three frames of bees left alive. Maybe crossing my fingers and feeding them week after week had made all the difference. Or maybe the bees would've had everything under control without any help from me. Even with all we've learned about honeybees, sometimes it's still hard to know just how things will turn out.

But caring for bees has always been a bittersweet endeavor. Now, thanks in part to the loss of natural habitat, threats from pesticides, and the specter of Colony Collapse Disorder, so is caring *about* them. Just as I'm never completely sure whether a bee season will be good or bad, I'm equally uncertain about

the future success of honeybees in general. Scour the unsettling reports about honeybees vanishing like Harry Houdini or see enough of those wide-eyed news anchors wondering, "Where have all the honeybees gone?" and it's easy to see the honey jar as half empty. (Sometimes I do catch myself thinking, *I should have held out for chickens. They're cuddly, productive, and as far as I can tell, they don't sting.*) Really, though, I wouldn't trade my bees for anything. And despite the mess we're presently in, I remain hopeful—and grateful. Grateful to have gotten to know my honeybees as well as I have. Grateful that much-needed time, attention, and funding are finally being funneled into research about them. And grateful, too, that people like you are coming to appreciate *Apis mellifera* for who she is—so much more than her honey and sting.

FURTHER READING AND RESOURCES

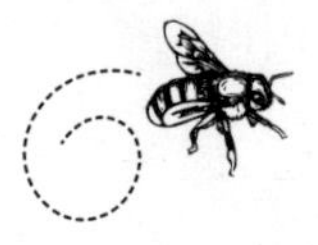

A. I. Root Company. *Bee Culture: The Magazine of American Beekeeping*, www.beeculture.com.

Crane, Eva. *The World History of Beekeeping and Honey Hunting*. Routledge, 1999.

Dadant and Sons. *American Bee Journal*, www.dadant.com/journal.

Flottum, Kim. *The Backyard Beekeeper: An Absolute Beginner's Guide to Keeping Bees in Your Yard and Garden*. Quarry Books, 2005.

Horn, Tammy. *Bees in America: How the Honey Bee Shaped a Nation*. University Press of Kentucky, 2006.

Hubbell, Sue. *A Book of Bees, and How to Keep Them*. Mariner Books, 1998.

Langstroth, L. L. *Langstroth's Hive and the Honey-Bee: The Classic Beekeeper's Manual*. Dover Publications, 2004.

Longgood, William. *The Queen Must Die: And Other Affairs of Bees and Men*. W. W. Norton & Company, 1988.

Maeterlinck, Maurice. *The Life of the Bee* [*La Vie des Abeilles*]. Dover Publications, 2006.

Shepherd, Matthew, Stephen L. Buchmann, Mace Vaughan, and Scott Hoffman Black. *Pollinator Conservation Handbook: A Guide to*

Understanding, Protecting, and Providing Habitat for Native Pollinator Insects. Xerces Society, 2003.

Shimanuki, Hachiro. *The ABC and XYZ of Beekeeping.* A. I. Root Company, 41st edition, 2007.

Winston, Mark L. *The Biology of the Honey Bee.* Harvard University Press, 1991.

For more information on beekeeping and bees, contact:

The British Beekeeping Association
The National Beekeeping Centre
National Agricultural Centre
Stoneleigh Park, Warwickshire CV8 2LG
www.britishbee.org.uk

The Save Our Bees Campaign
www.saveourbees.org.uk

National Bee Supplies
Merrivale Road, Exeter Road Industrial Estate,
Okehampton, Devon EX20 1UD
www.beekeeping.co.uk

The Xerces Society for Invertebrate Conservation
4828 SE Hawthorne Blvd.
Portland, OR 97215
(503) 232-6639
www.xerces.org

PERMISSIONS

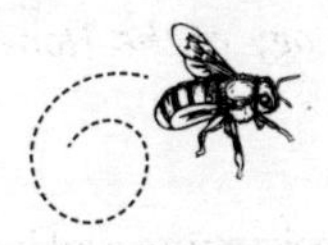

IMAGES

Pages 9, 24, and 34: From *The Biology of the Honey Bee*, by Mark L. Winston, pages 9, 156, and 209, Cambridge, Mass.: Harvard University Press. Copyright © 1987 by Mark L. Winston. These line drawings are reprinted by permission of the publisher.

Page 38: From *Honey Bee Biology and Beekeeping*, by Dewey M. Caron, Wicwas Press, LLC, 1999. Photos reprinted by permission of the publisher.

Page 125: From *Fuzzy Logic*, by Keith Bellinger, Farm & Garden online (www.farm-garden.com). Photo reprinted by permission of the author.

TEXT

Pages 9, 23–24, 34: From *The Biology of the Honey Bee*, by Mark L. Winston, pages 9, 156, and 209, Cambridge, Mass.: Harvard University Press. Copyright © 1987 by Mark L. Winston. Excerpt reprinted by permission of the publisher.

Pages 66–67: From "Man Wears Bees," *Life Magazine* 21 (September 9, 1946). Excerpt reprinted by permission of the publisher.

Pages 124–127: From *Fuzzy Logic*, by Keith Bellinger, Farm & Garden online (www.farm-garden.com). Photo and text reprinted by permission of the author.

Pages 142–143: From *The World History of Beekeeping and Honey Hunting*, by Eva Crane, Routledge, 1999. Reprinted by permission of the publisher.

Pages 146–150: All recipes are reprinted courtesy of the National Honey Board (www.honey.com).

Page 156: From *Madame Tussaud: A Life in Wax*, by Kate Berridge, New York: HarperCollins. Copyright © 2006 by Kate Berridge.

Pages 163–164: "Nancy's Lip Balm" from *The Backyard Beekeeper: An Absolute Beginner's Guide to Keeping Bees in Your Yard and Garden*, by Kim Flottum, Quarry Books, 2005. Reprinted by permission of the publisher.

ACKNOWLEDGMENTS

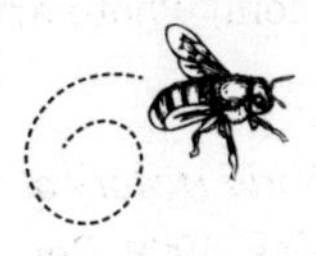

Special thanks go to John Ambrose, Jerry Bromenshenk, Dewey Caron, Norm Gary, Morfy Gikas, Jeffrey Harris, Tammy Horn, Tracy Hunter, Gene Kritsky, Gene Robinson, Matthew Shepherd, and Bruce Wolk for their expertise and generosity of time.

I am also grateful to Tina Carvahlo at the University of Hawaii, the Dartmouth Electron Microscope facility, Rick Dietz, Kim Flottum, and James E. Tew for allowing me access to their remarkable photos and to Keith Bellinger for his first-rate bumblebee house plans.

Thanks also to Marian Lizzi and Debra Goldstein for their encouragement and guidance and to Christina Lundy for her careful attention to detail.

Finally, I must mention Michael White. Thank goodness for his keen, garage-saling eye, without which I might not be a beekeeper at all.

INDEX

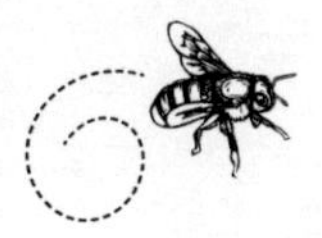

Page numbers in *italic* indicate illustrations; those followed by "n" indicate notes.